INCREASING PRODUCTION FROM THE LAND

A Source Book on Agriculture for Teachers and Students in East Africa

Andrew Coulson, Antony Ellman
and Emmanuel Mbiha

MKUKI NA NYOTA
DAR—ES—SALAAM

Published by
Mkuki na Nyota Publishers Ltd
P. O. Box 4246
Dar es Salaam, Tanzania
www.mkukinanyota.com

ISBN 978-9987-08-356-5

Distributed worldwide outside Africa by African Books Collective.
www.africanbookscollective.com

For the small farmers of East Africa,
and all who support them

Contents

Acknowledgements vii
The Authors. ix
Foreword xi
Abbreviations and Acronyms. xv

INTRODUCTION
Why this book, and how to use it 1

PART 1   A GUIDE TO AGRICULTURAL PRODUCTION 7
Chapter 1   How Plants Grow, and What They Need
To Grow Well. 9
Case Study 1: How Villages in North-East Tanzania
Can Adapt to Climate Change 22
Chapter 2   Crops and Livestock: How People Use the Land 27
Case Study 2: Sweet Potatoes – the Perfect Crop?. 47
Chapter 3   Water and Irrigation 51
Case Study 3: Dakawa Rice Farm. 67
Chapter 4   Agricultural Research. 71
Case Study 4: Research and Development of Cassava
in Tanzania 86

PART 2   GETTING THE MOST FROM THE LAND: HOW TO
RAISE AGRICULTURAL PRODUCTIVITY. 93
Chapter 5   Small Farms 95
Case Study 5: Profitability Analysis of Small Farms in
Northern Tanzania. 111
Chapter 6   Large Farms 115
Case Study 6: Upper Kitete Co-operative Farm. 128

Chapter 7 Market Access and Value Chains 136
Case Study 7: Tanzania's Co-operatives Look to the Future. . .152
Case Study 8: Cafédirect – Fair-Traded Tea from Tanzania
and Uganda. 158
Chapter 8 Resources for Agriculture – Credit and Contracts. .164
Case Study 9: Credit and Contracts – Contrasting
Experiences with Cotton and Tobacco in Tanzania 177
Case Study 10: Dairy Quality Chain – Tanga Fresh Limited. .181
Case Study 11: Contract Cultivation of the Antimalarial
Plant *Artemisia annua*. 185
Chapter 9 Agricultural Extension: Getting New
Technologies to Where They are Needed. 192
Case Study 12: Cashew Nuts in Vietnam and Tanzania 203
Case Study 13: Potatoes in Njombe: Dissemination
Without Extension. 204

PART 3 PRACTICE AND POLICIES 209
Chapter 10 Purchased Chemicals, Genetically Modified
Seeds – and the Alternatives 211
Case Study 14: Data on the Profitability of Organic Farms . . .225
Chapter 11 Gender Myths and Half-Truths 230
Case Study 15: Labour Allocation of Women,
Men and Children in Tanzania 241
Chapter 12 Agricultural Policies. 245
End game: A short quiz. 260

INDEX 272

Acknowledgements

Writing a book with a broad canvas involves far more people than its authors.

Many of our colleagues made detailed comments or suggestions as the work proceeded, though probably none of them will be entirely happy with what has resulted. Ben Wisner enabled us to use material from a team of researchers who worked on the impact of climate change in north-east Tanzania. The International Potato Centre gave us permission to use an extract about sweet potatoes from their website. Jeremy Jones provided guidance on irrigation, genetic modification, and matters of chemistry and biology. Emmanuel Sulle gave us advice on large scale land holdings. Arlene Inocencio gave us permission to use data from an International Water Management Institute publication about irrigation. Anna Mdee made it possible for us to use the study of the irrigation project at Dakawa and advised us on organic farming, as did Janet Maro and Alex Wostry of Sustainable Agriculture Tanzania, in Morogoro. Catherine Njuguna of IITA in Dar es Salaam provided expert assistance for the case study of cassava. James Curry Publishers allowed us to use material on contract farming of tobacco from the collection *Tanzanian Development: A Comparative Perspective* edited by David Potts. Frédéric Kilcher shared his work on Farmer Field Schools and helped us turn it into the case study of the motivations of small farmers. He also wrote the quiz that concludes the book. The International Labour Organisation gave us permission to use an extract from a report by Andrew Bibby on the future of co-operatives in Tanzania. Derek Byerlee of the World Bank in Washington was happy for us to use the quotation from a World Bank report. Chie Miyashita gave permission for us to use data on organic farming from his Master's dissertation. Romanus Dimoso encouraged us to use some of the data in his PhD dissertation, on the hours of work spent in the fields and in collecting water and firewood. Vicensia Shule advised us on matters connected with gender. We must thank Professor Damien Gabagambi for encouraging us to develop teaching material using local case studies.

Walter and Mkuki Bgoya and friends at Mkuki na Nyota helped to get the book through the publication process as quickly and effectively as possible. We also have a special debt to our copyeditor, Biddy Greene, who helped us to clarify what we wanted to say, and improved the style and consistency of the whole book.

Writing a book is slow and nerve-wracking, so we must also thank our wives and families for support and toleration during the process. The same goes for the many friends – too many to list separately – who encouraged us to keep going and finish the project.

Last but not least we could not have managed this without the many Tanzanian farmers we met over the years, or who contributed to the research projects that have shed light on what is involved with rural development in tropical Africa. This book is for them, with thanks, and for their children and grandchildren, in hope.

The Authors

Andrew Coulson worked as an economist in the Ministry of Agriculture, Food and Cooperatives in Dar es Salaam from 1967 to 1971 and taught economics and agricultural economics at the University of Dar es Salaam from 1972 to 1976. He taught at the University of Bradford, England from 1976 to 1982, and at Birmingham University from 1984, where he specialised in matters relating to local government in the UK. Since 2009 he has made regular visits to Tanzania, and published a range of articles relating to agriculture in Tanzania. A second edition of his book *Tanzania: A Political Economy* was published in 2013. He is current Chair of the UK branch of the Britain Tanzania Society. You can contact Andrew at a.c.coulson@bham.ac.uk

Antony Ellman is an agronomist and socio-economist. He first worked in Tanzania from 1962 to 1970, as manager of the co-operative farm at Upper Kitete, near Arusha, then as planner and adviser on smallholder development in the Ministry of Lands, Settlement and Rural Development, based in Dar es Salaam, but working in every region of the country. He spent the next twenty-five years doing similar work in Kenya, Uganda, Ethiopia, Sri Lanka and many other countries of Africa, Asia, South Pacific and the Caribbean, before returning to Tanzania in the mid-1990s to establish a smallholder tea and forestry programme in the Usambara Mountains. Since then he has undertaken consultancies in Tanzania and elsewhere on conservation agriculture, fair trade, and linking farmers to markets, particularly for the antimalarial plant *Artemisia annua*. Three of the case studies in this volume, and many of the proposals made elsewhere in the book, are based on Antony's work in Tanzania. You can contact Antony at antonyellman25@gmail.com

Emmanuel Reuben Mbiha was born in Kasulu, Tanzania. He is Associate Professor of Agricultural Economics at Sokoine University of Agriculture, Morogoro, Tanzania and was Head of the Department of Agricultural Economics and Agribusiness from 2000 to 2005. He was awarded a PhD in Agricultural Economics from Wye College, University of London (now part of Imperial College) in 1993. He started work as an Agricultural Extension Officer, later moving on to an academic position in agricultural economics. He has wide-ranging research experience in agricultural economics, post-graduate research guidance and community outreach. His major areas of interest include agriculture marketing and agricultural price and policy analysis. You can contact Emmanuel at Mbiha@sua.ac.tz

Foreword

This is a book about small farmers in Africa, because they are a reality and are likely to remain so for some time to come. It addresses the potential that small farmers have to get more out of the land, and to earn decent incomes while contributing to national development. It provides an introduction to the chemistry of soils and plant growth, the different ways in which the land can be cultivated and used, the special challenges of irrigation, and different kinds of agricultural technology and research. It describes the principles underlying soil management, crop and livestock production and how these principles are applied in practice by farmers and livestock keepers in East Africa.

It relates this knowledge to how people use different kinds of land, and the opportunities each type offers for the cultivation of crops and raising of livestock in East Africa. It gives special attention to livestock, both as a contribution to agriculture alongside crops (mixed farming) as well as a farming activity in its own right (pastoralism and ranching), putting all these activities in the context of climatic and demographic changes.

It is argued that the first essential for ensuring that land is productively used, is that farmers have some form of secure tenure which gives them an incentive to make long-term improvements and to use the land sustainably to cope with climatic change, soil degradation and demographic changes. In this regard, production levels must be raised while natural resources are conserved and not depleted or destroyed.

It is demonstrated that irrigation plays an important part in agriculture but it is easy to overestimate the benefits and to underestimate the associated technical challenges, competing demands and social challenges, especially getting groups of farmers to work together to share the water that becomes available and to take responsibility for the maintenance of the irrigation schemes.

In examining agricultural technology this book observes that much that is proposed to smallholder farmers in Africa is not acceptable to them. It makes a case for understanding the problems that farmers are facing, and for developing technology that is suitable for their needs by first looking at what they themselves are experimenting with, and then using this as a starting point for joint research and development by farmers and scientists.

The book addresses several myths and misunderstandings about small farmers. It starts from first principles explaining how plants grow and what is needed to make them grow better, addresses smallholder farmers as people who are out to maximise their incomes and minimise their risks, and demonstrates the importance of markets and value chains. Many marketing opportunities for small farmers are influenced by institutions and organisations (in the public and private sector) involved in marketing. It is demonstrated that analysis of value chains is the best way to get more value to farmers, as well as investing money off the farm, for example through improvements to storage, transport infrastructure, better packaging, and marketing. An important aspect of marketing is access to marketing information on price levels at different points in the marketing chain or in different places. The book examines the historical development of co-operatives, and sees a role for them in the future, provided they are facilitated to undergo transformation into viable businesses.

The book addresses three categories of schemes that are a means through which farmers hope to get higher prices for the crops they sell. These are agricultural credit, contract farming (which gives farmers guaranteed markets) and warehouse receipt schemes.

Access to credit and other forms of finance to small farmers is crucial, but presupposes availability of markets. Banks adopt various strategies to minimise the risk of lending, by lending to groups of farmers, co-operatives or through contract farming. The book addresses the opportunities for transfer of agricultural technologies and information that can help farmers increase their production, including an assessment of extension services.

It is argued and well demonstrated that farming is a business in which small farmers make choices with the resources (land, labour and capital) available to them. They make decisions about which crops to plant, where and how much should be planted, at what times, and using which seeds and inputs. These are business decisions in which farmers' local knowledge is of great value. It enables them to engage in innovations as they try out different ways of producing what they need, and learn from experience what works best. It is shown that what farmers do is governed by the desire to secure their food supply for the forthcoming dry season and to minimise the risks of failure. In this context, many farmers manage risk by engaging in mixed farming and by earning incomes from outside their own farms as providers of seasonal labour or producers of non-farm outputs.

The dissemination of an innovation may not need extension workers at all if market forces operate to demonstrate the obvious benefits from the innovation - such as the spread of potato farming in Njombe where farmers copied from each other. The book demonstrates other examples of innovations that spread through market forces. A process of agricultural innovation may be assisted by mass media that describe successful new activities. The key to understanding extension is to realise that it may be organised in different ways. The conventional extension service approach is criticised for being top down, and alternative approaches are discussed such as identifying innovating farmers in an area, and then taking steps to diffuse their innovations widely. Specialist research workers would be in the background, available as consultants ready to make useful suggestions.

The book addresses practices and policies such as the opportunities and risks associated with a Green Revolution and with genetically modified (GM) materials, the position of women, and agricultural policies and the associated political economy analysis, including the powerful interest groups which may be in conflict with the farmers' best long-term interests. The book describes the issues raised by genetic modification, and suggests alternatives that are less environmentally and socially damaging. It also addresses gender and what it may take to enable more women to develop their talents in agriculture and elsewhere.

This book finally brings together some of the key themes discussed, and summarises the policies that are needed to promote a more productive, and sustainable, agriculture. The messages presented are useful to those in governments, donor agencies, NGOs, farmer organisations and investors who are able to influence policies that relate to agriculture in Africa.

The book provides a valuable guide for teachers and students of agricultural and rural development by setting out key issues and providing illustrations and case studies based on local experiences. It is relevant to the work of central and local governments as well as non-governmental organisations involved in promoting rural and community development. Most of the examples are from Tanzania, but the principles are largely applicable across tropical Africa.

Each of the twelve chapters starts with lists of key themes or concepts to be discussed, and concludes with short case studies which illustrate some of the principles set out in that chapter. Following that is a list of material for further reading or study. Each chapter ends with a set of exam or essay questions which explore the key issues.

Professor Sam Wangwe
Executive Director, Economic and Social Research Foundation, Dar es Salaam, 1994–2002, and of REPOA 2011–2016

Abbreviations and Acronyms

AAL	African Artemisia Ltd.
AGRA	Alliance for a Green Revolution in Africa
CA	conservation agriculture
CGIAR	(formerly) Consultative Group for International Agricultural Research
CIP	International Potato Center (Peru)
DEFRA	Department for Environment, Food & Rural Affairs (UK)
FAO	Food and Agriculture Organization of the United Nations
FLO	Fairtrade Labelling Organisation
GM	genetically modified
ha	hectare(s)
IITA	International Institute of Tropical Agriculture
MEMCOOP	Member Empowerment and Enterprise Development Programme
NAFCO	National Agricultural and Food Corporation (Tanzania)
OFSP	orange-fleshed sweet potato
PCS	primary co-operative society
REPOA	(formerly) Research into Poverty Alleviation, Dar es Salaam
TDCU	Tanga Dairy Co-operative Union
TFL	Tanga Fresh Ltd.
WEMA	Water Efficient Maize for Africa

MEASURE CONVERSIONS
1 hectare ≈ 2.47 acres; 1 acre ≈ 0.4 hectares

INTRODUCTION

Why this Book, and How to Use it

This is a book about small farmers in Africa.

Why *small* farmers? Above all, because they are a reality and will remain so for at least a generation. But also because, while they make big contributions in all African countries, they could do even more.

In the formal sectors – where there are regular wages and contracts of employment – the number of jobs being created are far fewer than the numbers of young people leaving schools and looking for work. So only a lucky few are employed on acceptable terms. Most survive with insecure and poorly paid work in the informal sectors in cities and towns, or in the rural areas. Others remain dependents of their families well into adulthood.

In the rural areas most families have access to land. They grow as much of their own food as they can, because purchasing food is expensive and if they have land they can grow food with minimal cost.

Rural households also earn income from selling other products. Agriculture has advantages over most other ways of earning cash. It is a means of survival in which the skills are, in broad terms, well known, and the risks are less than in most alternatives. Farmers mostly sell surpluses of crops grown for food, or livestock or products derived from livestock – such as milk or eggs. Some sell products for export that were introduced to the country in colonial times – such as cotton, coffee or tea. Farmers who can access urban markets, or who live in urban areas and have access to land, may grow and sell vegetables or fruits to the urban population.

For well over a hundred years, governments in Africa have been trying to persuade farmers to extend the areas of land they cultivate, and to use mechanisation, or purchased inputs such as fertilisers or insecticides, to get greater yields from small areas. In this they have had some success. But while agriculture is less risky and uncertain than many other business activities, there is still much that can go wrong: for example, failures in the rains, attacks by pests or plant diseases, or unexpectedly low market prices. So most small farmers are cautious. If the outlay cost is low, a good return likely, and inputs are available free or on credit, many will try out new crops or alternative methods. But when there is a risk of failure, or doubts about whether they will be paid, farmers are understandably very reluctant to purchase inputs.

Overall, farming outcomes in recent years have been good. For example in Tanzania between 1988 and 2012, the value of agricultural production, adjusted to remove the effects of inflation, grew on average at more than 4% per annum [1, Table 7.2]. For agriculture, in countries around the world, such sustained growth rates over many years are very unusual. The growth rate was above the rate of population growth, and was achieved despite the challenges posed by climate change and the long-term declines in the prices on world markets of many of Tanzania's traditional export crops, such as cotton, coffee and tea. Most of the growth was in crops grown for food: mainly maize and rice, but also potatoes, sunflower, tomatoes and other vegetables. Much of it came from small farms. It is an achievement which should be celebrated.

A recent report [2[suggests that there has been a remarkable rise in the number of larger farms – of 10 to 100 hectares – in Ghana, Kenya, Tanzania and Zambia. Some of these farms were created by existing farmers who took over vacant land or land previously farmed by small farmers who for one reason or another were not able to carry on. Many are on land of low quality, previously used mainly for grazing. But many of the new larger farmers have made money through employment in the public or the private sectors, and believe that they can make money by investing in agriculture. This too is a reason for celebration.

Large farms in Africa face many challenges, however. The owners need to employ labourers: either for short periods – with problems of supervision and training – or they must keep permanent workers employed throughout the year. They also have initial capital costs for developing the land, and then on-going overhead costs. They risk soil erosion, especially if they use tractors. There have also been many failures with markets. Moreover, if too much land is taken over by large farms – including common land which was previously available for grazing, collection of fodder or fuelwood – small farmers will find it much harder, sometimes even impossible, to survive, and this will add to the extent of rural poverty, or, if people who become landless are obliged to move to towns, to urban unemployment and the social problems that can result.

Large scale farming has its place, but it is not the only way ahead. The potential exists for small farmers to get much more out of the land, without undue risk, and to earn good incomes, while contributing to national development. This book sets out how this can be done.

MYTHS AND MISUNDERSTANDINGS

Many statements about agriculture in Africa are either misleading or not supported by evidence. One myth has just been discussed – that small scale agriculture has no medium- or long term future.

Other myths or half-truths are discussed later in the book. Examples of such 'myths' are: that there is unused land that is suitable for agriculture, so anyone who wants land can have as much as they desire; that there is plenty of water, giving great scope for increases in irrigation; that small farmers do not innovate; that livestock economies, especially pastoralism, are unproductive; that small farmers are stubborn and irrational, and that there is no logic in what they do; that men are more productive in their agricultural activities than women. None of the above are correct in all circumstances. We discuss some of these myths in Chapters 5 and 11, and show that they are not based on fact.

THE ARGUMENT OF THE BOOK

To address these myths and misunderstandings, this book starts from first principles. Part 1, four chapters, explains how plants grow and what is needed to make them grow better. The first chapter is about soils, nutrients, water and the chemicals that plants must have to grow. The next chapter shows how human beings have adapted what nature has given, to feed an ever-growing world population. It shows that agriculture is not straightforward: a machine can process one product into another with almost complete reliability; with any form of agriculture, little can be predicted accurately, and much can fail, often very suddenly. Agriculture is a business, and almost as much an art as a science; it depends on hunches and judgements, and on the minimisation of risks.

The third chapter is about water – which is fundamental to all forms of life – and its use in irrigating crops. There is optimism about irrigation in Africa, and yet in the past many large-scale irrigation schemes have failed. This chapter sets out the choices in irrigation, and describes what is needed to make it a success.

The fourth chapter is about agricultural technology and research. It describes how agriculture can become more productive, and explains some of the difficulties and risks that can occur.

Part 2 of the book, five chapters, is about people. Chapter 5 is about small farmers, what motivates them, and their potential to generate surpluses of agricultural products. It draws on the work of Frédéric Kilcher who has worked with groups of small farmers in Tanzania to help them maximise their incomes and minimise their risks.

Chapter 6 compares small-scale farming with farming on a larger scale. It describes the situations in which large-scale farming is likely to succeed, as well as its limitations and costs.

The next two chapters are about marketing. This is key to farming success; if marketing is not efficient, crops will not be sold, or prices paid to farmers will be lower than necessary. The first of these chapters introduces the concept of a value chain for each crop, identifying where the costs of marketing are highest, and showing how they may be reduced. It then looks at the main institutions involved in marketing, in both private and public sectors. The following chapter is about credit and other forms of finance for small farms. Credit will not be available unless markets are reliable, and even then a bank or other financial institution that lends to farmers may not get its money back. Banks may be able to reduce their risks by lending to groups of farmers, or to co-operatives, rather than to individuals. Another approach is through "contract farming", where a farmer receives support from a processing company or marketing agent and signs a contract to supply that agent or company with crops of a specified quality at a specified time. Contracts of this kind are the basis of much of the farming in developed countries; but they can also turn sour – for example if the farmers are not paid as agreed or expected, or if other parts of the contract are not delivered.

The last chapter in this section shows how innovations may be communicated and spread. This does not necessarily involve government-employed agricultural extension workers: many innovations are simply copied by one farmer from another.

Part 3, about practice and policies, has three chapters. Chapter 10 is about the opportunities and risks associated with the so-called Green

Revolution and with genetically modified (GM) materials – including the risks to those who use purchased chemicals, and to the environment. Chapter 11 shows how women are often exploited, or marginalised, even though they do much of the hard work, and are used to making difficult decisions. They have the potential to do more, but their rights need to be protected; men have to accept that women have much to offer.

The final chapter is about agricultural policies. It is addressed to political leaders and administrators, who have to make choices and decisions about what to support, and to donors and NGOs. But it is also addressed to farmers themselves, especially small farmers, who need the confidence to stand up to powerful interest groups who may propose solutions which are not in the farmers' best long-term interests.

HOW TO USE THIS BOOK

This book is in the form of a textbook, but it is not a text for a particular course. It can better be seen as a sourcebook, or guide, providing a context, or background material, for anyone teaching or studying rural development. More than fifty universities and university colleges in Tanzania alone, and many colleges offering diplomas or other qualifications, teach courses which include material about rural development. This book can assist both teachers and students by setting out the issues and providing illustrations and case studies based on local experiences. It can also help those whose work in central or local government and in non-governmental organisations involves promoting rural and community development.

Most of the examples are from Tanzania, but the principles do not change across tropical Africa, so the book has a much wider application. We hope that people in other African countries will find it useful too.

The twelve chapters each start with lists of key themes or concepts to be discussed, and conclude with a short case study which illustrates some of the principles set out in that chapter. Following that is a list of material for further reading or study, all of which is available on open access. Anyone with access to the internet can copy the link and download the article (in some cases this is a whole book). If you find it difficult to copy out the whole web-address for an article, it may be easier to put the title and author into a search on Google and find it that way. But we can confirm that these articles are on the internet, because we have ourselves downloaded them. For a few chapters, including this Introduction, there are also a small number of references which are not on open access.

Each chapter ends with a set of exam or essay questions which explore the key issues. There is an index at the end, so that anyone interested in a particular topic can quickly find the places where it is discussed. The emphasis throughout is on making the book easy to use, making each chapter more or less able to be read on its own, and providing practical help for students and their teachers.

Readers can use the book in different ways. Some will read it through from the beginning to the end. Those familiar with a topic can skip to a later point (though some may be interested in how a familiar topic is treated). Others can use the index to find what is said about a particular topic.

If you find this book useful, please contact us and tell us if there is anything that you think we should have added – or developed in more depth. If you do not agree with our approach, again please tell us, and, if there is a second edition, we will carefully consider your points.

There is a lot at stake. In Tanzania alone, at least thirty million people in rural areas are active in farming, even though for many this is not their only source of income. Many will migrate to urban areas, but population growth will mean that the numbers in rural areas are unlikely to fall for many years. Climate change means that the challenges of using land productively and sustainably are likely to become more complex.

If this book helps those who read it to understand the challenges that face small farmers, and how they can overcome them, it will have achieved its objective. The potential for much-increased production and better livelihoods is there.

REFERENCES

1. Christopher Adam, Paul Collier and Benno Ndulu (editors) *Tanzania: The Path to Prosperity*. Oxford University Press, 2017
2. Thomas Jayne, Jordan Chamberlin, Lulama Traub, Nicholas Sitko, Milu Muyanga, Felix K. Yeboah, Ward Anseeuw, Antony Chapoto, Ayala Wineman, Chewe Nkonde, Richard Kachule "Africa's changing farm size distribution patterns: the rise of medium-scale farms" *Agricultural Economics* Vol. 47, 2016, pp.197–204.

PART 1

A GUIDE TO AGRICULTURAL PRODUCTION

The first section of this book provides an introduction to agricultural technologies – the chemistry of soils and plant growth, the different ways in which the land can be cultivated and used, the special challenges of irrigation, and different kinds of agricultural technology and research.

CHAPTER 1

How Plants Grow, and What They Need To Grow Well

Key themes or concepts discussed in this chapter

- How plants grow by using chemicals from the soil and the air.
- The importance of the soil.
- Why water is so important for plants, and how it can be conserved.
- How the sun provides the energy that makes plant growth possible – and how too much sun makes this increasingly difficult.
- How chemicals that are lacking can be supplemented by organic manures, or by synthetic fertilisers produced in factories.
- Why there are more plant and animal pests and diseases in Africa than anywhere else in the world, and what can be done to combat them.
- The importance of maintaining soil structure – the advantages and risks of mechanisation.
- Soil erosion, and steps that can be taken to prevent it.
- The context for a discussion of agriculture in Africa – the likely impacts of global warming and soil degradation.

HOW PLANTS GROW

Agriculture is about how soils can be used by people to grow crops or to raise livestock.

Plants grow through chemical processes in which they take energy from the sun, carbon dioxide from the air, mostly through their leaves, and water and other chemicals from the soil through their roots. The scientific name for using the sun's energy is *photosynthesis* – literally "making things from light".

Animals, including insects etc., grow by eating material from plants or other animals.

Plant growth is a process of copying. The blueprints, or plans, are *genes*. Every piece of living material has its unique *genetic code*, which determines which chemicals go where in order for the organism to grow. If a plant is grown entirely from an existing plant, as when a shoot from a tea bush or a stem from a cassava plant is put in the ground and starts to grow, its genes will be the same as those of the original; and if the environment is the same it will grow to be just like the original. This is called *vegetative propagation*.

However, most plants are the product of two sets of genes – male and female – usually coming from two different plants of the same species. In this case the new plant is a mix or a *hybrid*; it is not an exact copy of either original. As described in a subsequent chapter, this is the basis of plant and animal breeding – plants which have desirable properties, perhaps in terms of yield per plant, or in resistance to drought or disease, are identified from the mix available and are used to breed further plants. Similarly, the best animals are kept for breeding and some of their best characteristics are inherited by their children. And in this way the genes – and hence the plants or animals – are modified and improved.

Genes may also be modified in a laboratory. Genetic material from other plants or organisms may be added to the genes in a plant or animal, or existing genetic material may be removed or "knocked out". This is called *genetic modification*, or *genetic engineering*. How it is done is explained in Chapter 10, along with the wider issues it raises.

The most important chemical in plant growth is *carbon*. Carbon is the basis of all *organic compounds*, and the study of these compounds is called *organic chemistry*. Organic compounds include *carbohydrates, fats* and *proteins* – the basic foods of animals. They also include hydrocarbons, the basic components of coal, oil and natural gas. Carbon is also a key element in most drugs and dyes. The leaves of plants have the ability to take the gas carbon dioxide from the air and, with water, to turn it into carbohydrates. These are then distributed to other parts of the plant, such as the stems and roots, and turned into other organic products. Carbon dioxide is what is known as a *greenhouse gas*. It absorbs heat radiated from the surface of the earth, making the earth warmer than it would otherwise be. In other words, carbon dioxide is one of the main causes of *global warming*, and plants which take in carbon dioxide through their leaves are one of the main ways of combating it. There is more about global warming at the end of this chapter.

Plants also need *nitrogen*, which is an important component of leaves. Nitrogen is one of the main components of the air we breathe, but in that form it cannot easily be absorbed by plants. But *nitrates*, which include nitrogen, are an important component of good soils. When leafy crops such as maize or cabbages are grown, nitrates are taken out of the soil. If the soil is losing nitrogen faster than it is being replenished, some way must be found to replace it. If this is not done, the soil will gradually lose its fertility and the yields of crops will go down.

There are five ways in which nitrogen and other chemicals can be put back into soils:

1. some plants have bacteria (in nodules on their roots) which can "fix" nitrogen, or take it in from the air. The most common nitrogen-fixing plants are legumes (such as beans or peas, or fodder crops such as clover or lucerne[1]). Legumes are valuable in agriculture because they provide a valuable source of proteins for both humans and animals, at the same time improving the fertility of the soil;

2. from the waste products of animals – "farmyard manure" – or from bonemeal;

3. by ploughing-in crop residues or plants grown for the purpose somewhere else – "green manure";

4. by preparing "compost" from leaves and other parts of plants, and from waste products from food, which are left for a time in a place where they can rot, before being taken to the fields.

5. If any or all of the above are not possible in sufficient quantity, then to get good yields, it may be necessary to add commercial chemicals – so called "nitrogen fertilisers".

Two other chemical elements are needed in substantial quantities for plants to grow, especially for the formation of roots and stems. They are also major constituents of the bones of animals. *Phosphate* and

---

[1] Also called alfalfa. In this book we call it lucerne.

potash (phosphorus and potassium in forms that dissolve in water) occur naturally in most soils, coming from the rocks from which the soil was created, or from the decay of plants. They can be added in green manure, in bonemeal (the bones of animals, ground into a powder), or in inorganic fertilisers. The production of phosphate is dominated by three very large international companies which are, effectively, in a position to set world prices.

There are relatively small deposits of phosphates, along with nitrates and potash, in "guano" – the droppings of bats and seabirds, found in places where they have accumulated for thousands of years. Guano is the basis of fertilisers from Peru and Chile, and from the Minjingu deposits in Northern Tanzania.

Calcium and *magnesium* are important for animals: they are key components of their teeth and bones. They are also important to plants, in countering the acidity of soils. Soils may become acidic if high rainfall washes certain elements out of it, or if the rain itself is acidic, or if the soil is waterlogged and plant material breaks down in the absence of oxygen. Calcium, usually in the form of lime (calcium carbonate), may be added to soil to reduce its acidity.

Sulphur is needed for plant growth, as are tiny quantities of "trace elements" – iron, zinc, copper, manganese, boron, molybdenum, iodine, chlorine and cobalt.

All of the chemical elements discussed above can be stored in what we call the soil.

SOIL PROVIDES AN ENVIRONMENT WHERE THESE CHEMICALS ARE AVAILABLE

Soil is a thin layer on the surface of the land, sometimes only a few centimetres deep, seldom more than a few metres. It differs from exposed rock, where little can grow, and from sand and gravel in deserts and beaches where the particles are so big that water simply drains through. Few plants can grow under such conditions.

Soils include very small particles – formed when rocks are broken down, usually over millennia, by ice, water, wind, and fire, and through the activities of plants and other forms of life such as mosses and burrowing animals (especially worms). Soils made up of the tiniest particles, less than 0.002mm in diameter, are called *clays*. Those consisting of slightly larger particles are called *silts*. Those with the largest particles are *sands*. Clay particles are so small that in the absence of water they can set hard, and water cannot penetrate such soil. That is fine for the bricks of a house; it is not good if plants are to grow in the soil. Good quality soils, called *loams*, contain a mixture of sands, silts and clay.

Soils can be moved, or eroded – as further considered below. Silt and stones are carried by streams or rivers to the bottoms of valleys, or into river deltas, and deposited when the flow slows down. The resulting *alluvial soils* are usually very fertile because of the chemicals they contain and pick up on the way. But this good soil is lost from wherever it originated. Even worse, landslides and mudslides can have tragic consequences for the people living nearby, and for the land they cultivate.

Soil can also be moved by the wind. This is not good if topsoil is blown away. More positively, soil is moved by "digging animals" such as rats, or by insects such as ants or termites. These movements are very valuable for agriculture because they bring back to the surface chemical substances which have been washed down into the soil. Last but not least, soil and rocks can be moved by geothermal processes, most obviously in volcanos. Many of the best soils come from volcanos – such as the slopes of Mount Kilimanjaro, Meru or Rungwe in Tanzania.

There are also good soils where rivers deposit soil when they flood – for example in the Rufiji Valley – or where they have done so in the past, as with some of the "black cotton soils" in valley bottoms, which can be used for growing rice and other crops. Soils can also be pushed upwards by forces in the core of the earth, so that, for example, limestone from the shells and bones of fish, which was at the bottom of seas millions of years ago, can even end up at the top of mountains.

There is a further component of soil, which is fundamental to its role in plant nutrition. This is called the "organic content", or *humus*. It is derived, mostly, from plant material and animal waste. Tiny "micro-organisms" – there are millions in just a handful of good soil – feed on this material and break it down into its basic elements, into forms that can be absorbed by the roots of plants. Most of these organisms are bacteria, but fungi, worms and insects also break down plant material. Humus also helps soils to hold on to water.

A high humus content ensures:

- that the soil does not set hard (and so is easy to cultivate),
- that the soil can hold water for long enough for plants to grow, but not so much that it becomes waterlogged, and
- that the soil retains the necessary chemical elements discussed above.

Without humus there would be little life on our planet. High temperatures and rays from the sun dry out the soil, and in extreme cases burn away the humus. In tropical conditions most of the humus is just below the surface, where it is protected from the sun. Further down, most of the chemicals have been washed through the soil and there is less organic matter.

Inorganic fertilisers work best when there is a high humus content in the soil. That is because the humus holds water, and the inorganic fertilisers dissolve in this water. In soils with little humus many of the chemicals in the fertilisers drain through the soil and are lost to the plants. The chemicals may even contaminate drinking water supplies. But if soils are well looked after, with plenty of compost and green manure added, they may not need factory-produced fertilisers (see, for example, [2] and [3] in the list of Further Reading at the end of this Chapter which can be downloaded from the Internet, especially the article by Natasha Gilbert).

There are, of course, costs in using inorganic fertilisers, and risks to water supplies and fish stocks. These fertilisers, and alternatives to their use, are discussed in depth in Chapter 10.

Last but not least, there are many empty spaces in a good soil, and they are filled with air. This air is needed for the decomposition (breakdown) of plant material to take place (by oxidation). If there is little or no air, for example because the soil is waterlogged, or because it has been compacted by the use of heavy machinery, then decomposition can only occur by processes which release unwelcome chemicals (not least the greenhouse gas methane) and which often turn the soil acidic.

THE EFFECTS OF WATER, SUN AND CULTIVATION

Weather conditions all over the world are becoming more extreme. Average temperatures are rising, and deserts, not least the Sahara Desert, are increasing in size. Rainfall is increasing in some places, decreasing in others. Almost everywhere it is becoming even less predictable than in the past. There are many areas where much of the rain comes in short, heavy storms which cause flooding and wash away the soil. Then there can be long periods with no rain at all.

In some upland areas there is rain and mist through much of the year. In other upland areas, two separate rainy seasons bring, overall, large amounts of rain, which means that two crops can be grown in a year. But in large parts of Africa the average amount of rain is less than 80 centimetres a year, with much of this coming in a few large storms, after which the soil quickly dries or becomes hard. Once the soil is dry, agriculture is almost impossible.

Such storms can destroy crops in the field and create different forms of erosion: primarily *sheet erosion*, where soil runs off the surface of the land taking the topsoil with it, and *gully erosion*, where the water digs deep and removes large areas of subsoil as well as the topsoil.

Where the top soil is removed, or is thin and lacking in organic matter, most of the rainwater either runs off or quickly drains through to deep levels, with the result that most plants cannot survive the long, dry season.

Thin topsoil also means that the sun bakes the surface of the ground hard, making it difficult to cultivate. However when the rains come, and crops are planted, there is plenty of sun as well as water, so the plants grow quickly. This applies to cereal crops (such as maize and rice), legumes (such as beans and lucerne), vegetables (such as tomatoes and potatoes), and many other crops. But it also applies to weeds, which must then be controlled – by hand weeding, spraying, ploughing, or (ideally) by maintaining crop residues as a surface mulch which smothers the weeds and prevents them from growing in the first place.

CONTROL OF SOIL EROSION

Farmers need to protect their soils from the negative effects of sun and water. The simplest way to do this is to leave the soil surface bare for as short a time as possible, especially after ploughing and after harvest. This can be done by planting quick-growing crops, or by covering the land with mulch – plant material of any kind. Perennial grasses also protect the soil against erosion, allowing rainwater to sink slowly into the surface layer where it is available for plant growth.

Another way to protect the soil against erosion is to plant crops on ridges which run "along the contours", i.e. across the slope of the land. Water is held between the ridges and does not cause damage lower down the slope. The simple technique of ploughing along a contour rather than up and down the slope can often make an enormous difference – ploughing up and down invites gullies to form where the water runs down the slope. On land with low rainfall, soil may be protected by digging trenches along the contours, which fill with water when rain falls. On very hilly ground it may be possible to construct *terraces* in which flat land is created and protected by walls.

Another effective measure is to plant, or keep, trees on the land. Their roots hold the soil, reducing the risk of erosion. Trees provide density for other crops, and protection from winds. They also attract mists and clouds, which increase rainfall, and their roots help conserve water sources. The growth of trees in much of Africa is far more rapid than in colder climates: spruce or pine trees that would take fifty to seventy years to grow to maturity in Norway will be mature in the Southern Highlands of Tanzania in about twenty years.

Land that is close to springs and watercourses is often at risk of gulley erosion and should not be cultivated. Ideally such areas should be protected by the designation of forests in which cultivation is prohibited. Population pressure on the land, however, often leads to such conservation measures being ignored.

SHIFTING OR CONTINUOUS CULTIVATION

Most of the best soils in Africa are already in use for agriculture or other purposes. Much of this land has been cultivated repeatedly for many years and is starting to lack some of the important nutrients. Traditionally, in what is sometimes called "shifting cultivation", such land was rested for several years and allowed to regain its fertility. But with increasing population, and a move to live in villages, this is often no longer possible. The move to "continuous cropping", where the land is cultivated every year without a break, requires crop rotations, i.e. growing a different crop on each piece of land in successive years. In addition, as much animal or green manure as possible should be used, and an understanding is required of which minerals are lacking and need to be replaced. Continuous cropping is not straightforward – it is one of the fundamental challenges of agriculture in dry places.

PESTS AND DISEASES

There are probably more plant and animal pests and diseases in Africa than anywhere else on earth. There are good reasons for this. In temperate climates, long cold winters kill off many of the unwanted organisms, whereas in Africa they are much more likely to survive.

The pests include birds, rodents, armyworms, locusts, monkeys, wild pigs, elephants and hippopotamus. There are also many insects which attack crops, either in the field or in storage. These include *aphids* such as greenfly and blackfly (which suck nutrients from the plants), slugs and snails and caterpillars (which eat the leaves of plants), and stalk or stem borers, and mealybugs.

Ticks, mites and tsetse fly are three of the insects which cause diseases in animals, however this is not the place to describe every plant or animal disease. Those who want this information can look at Brigette Nyambo's survey in the Further Reading [4]. What follows is selective, to illustrate the wider picture. The issues which arise when powerful chemicals (insecticides) are used to kill insects are considered in Chapter 10, as are the risks involved when the genes of plants are altered by genetic modification so that they can resist insecticides.

Viruses. These are minute organisms – much too small to be seen even under a microscope. They are responsible for many human and animal diseases, which spread by direct contact of one sort or another, or via droplets in the air. Examples are HIV, rabies and smallpox, in humans; Newcastle disease in poultry; rinderpest and foot and mouth disease in cattle. Viruses carried by insects cause some of the main diseases of plants, for example, maize streak disease, cassava mosaic and brown streak diseases, tobacco mosaic, groundnut rosette disease, and sweet potato virus disease.

Bacteria. These are also tiny organisms, but they can usually be seen under a microscope. They cause many very unpleasant diseases in humans and animals, such as anthrax, tuberculosis and bovine tuberculosis, leprosy, and gonorrhoea. Some are carried on insects and damage plants, resulting for example in bacterial wilts and blights. But some bacteria can also be very helpful to plants, for example nitrogen-fixing bacteria on the roots of legumes, or bacteria that help to break down plant material into its constituent materials.

Fungi (or funguses, or moulds). Many of the most damaging plant diseases in Africa are a consequence of fungi, such as the fusarium wilts which attack maize, bananas and cotton, or wilts which attack the leaves or roots of cassava. As they grow these fungi release enzymes which break down plant material into basic products. They then absorb some of these products into their bodies, but they leave much more behind, which other plants can use. Fungi also cause diseases known as rusts, blights, leaf spot diseases and mildews (all occurring on cereals such as wheat, maize, rice). Aflatoxin is a toxic chemical produced by fungi, which is particularly common in groundnuts, and there are many other fungi, which are harmful either to the plants or to the people or animals who eat them.

Nematodes or small worms. These exist in the roots of plants, and can cause grievous harm, for example to bananas, to root crops such as carrots or potatoes, and to fruit trees.

CONTROLLING PLANT AND ANIMAL DISEASES

The traditional way of combating diseases in plants is to remove any affected material and destroy it by burning or burying it deep in the ground. The impact can also be minimised by planting plants in small plots, or far apart, so that diseases spread more slowly.

The near perfect way of combating diseases is to breed varieties that are resistant to the common diseases. Much plant breeding – discussed in Chapter 4 on agricultural research – has this as its aim. For example recently released varieties of cassava have improved resistance to cassava mosaic disease and its virulent "Uganda variant", which has discouraged the growing of the crop in many parts of East Africa. But plant breeding is a slow process, and cannot be done for every disease and for every different climatic condition.

Another near-perfect approach (when it works well) is to introduce new "biological controls" which attack the insects which carry the disease or damage the plants. Thus when cassava was introduced to Africa from South America, it came without its traditional pests, including the mealybug. But in the 1970s the mealybug reached Africa, and spread rapidly. In South America it is not a major problem because other insects keep it under control. So scientists at the International Institute of Tropical Agriculture (IITA) in Nigeria studied these insects, and eventually introduced one of them to Africa, a small wasp that fed on the mealybug. This has removed a major problem without farmers having to take any action. A study by Peter Neuenschwander [5] showed that it was a hugely cost-effective piece of research, since every grower of cassava benefited.

Not all "biological controls" of this kind have succeeded. For instance scientists have been trying for fifty years to control sleeping sickness, which is caused by tsetse fly, by introducing sterile male tsetse flies which would not be able to reproduce so would ultimately lead to the disappearance of the tsetse fly. They have not yet succeeded.

There are other non-chemical methods of controlling pests. Some plants, such as marigolds and onions, deter certain insects, and may be planted close to threatened crops. Or they may be planted around a crop to create a barrier. The Gatsby Foundation in Tanzania promotes various means of controlling pests by planting other crops which produce chemicals that deter insects. Article 6 in the Further Reading [6], about "push-pull farming", describes how a legume, *Desmodium*, was planted around a field of maize. The legume repelled some of the insects that affect maize and also reduced damage from the parasitic weed *Striga* which attaches itself to the stalks of plants and sucks their juices. It also provided useful feed for cattle.

Insects and funguses can also be controlled by the use of commercially produced chemicals – often extremely unpleasant and dangerous chemicals, poisonous to animals and humans as well as to insects. They are sprayed on to crops – with a risk that some of the spray might get into the lungs of the operator, or into the food chain when the crops are eaten. They are also, in many cases, expensive. Most large-scale agriculture is dependent on this kind of chemical control, but there are many possible downsides and risks. These issues are further considered in Chapter 6 and especially in Chapter 10.

CLIMATE CHANGE AND GLOBAL WARMING

The final section of this chapter puts what has been written above into the wider context of what is happening to agriculture in the world, especially in Africa.

The overwhelming majority of scientists do not doubt that the planet is getting warmer. The polar ice caps are breaking up; glaciers are melting from the Himalayas to the Alps to Mount Kilimanjaro; the Great Barrier Reef in Australia is losing its coral; there are more big storms than in the past; and record temperatures – both high and low – are being recorded in almost all parts of the world. When temperatures are too high, soils dry out, plants burn up, and in extreme cases formerly productive areas turn into deserts.

Almost all scientists agree that the main cause of this warming is human activity, in particular the discharge of carbon dioxide, methane and other "greenhouse gases" into the atmosphere. There is no doubt that the quantities of these gases in the air have risen, roughly alongside the use of fossil fuels, and there is strong evidence that this is a cause of the warming. If this pollution of the atmosphere continues, there is every indication that average temperatures around the world will rise by at least 2 to 3 degrees Celsius, and average sea levels will rise by at least a metre – making some islands uninhabitable and requiring countries with low-lying coastal areas to spend huge sums of money on flood protection.

The effects of higher temperatures on rainfall are complicated. On the one hand, higher temperatures mean more evaporation from the sea, lakes and land areas, and hence more rain. But, on the other hand, where and how this rain falls will be hard to predict, and much of it will be in the form of heavy, violent and destructive storms and hurricanes. The increase in violent storms has been very noticeable in recent years.

Higher temperatures and more rain (even if unpredictable) is not bad for everyone. The climates of Siberia, the UK, and parts of Africa may become more pleasant and productive. Thus in the last ten years, most parts of Tanzania have had, on average, more rain than would have been expected from long term trends, and this has made possible higher production of maize and rice. However, some mountain areas in Tanzania, such as Kilimanjaro, the Pare and Usambara Mountains, and Kagera, have had less rain, with bad consequences for their agriculture. In Uzbekistan, the Aral Sea is now less than a tenth of its size only a few years ago, and is highly polluted as well – partly a consequence of over-abstraction of water for industry and irrigation. In West Africa, lower flows in the Niger River (a consequence of lower rainfall and increased irrigation) have reduced Lake Chad to a small fraction of what it was, with significant consequent losses of agricultural production and fisheries. Many farmers in the region can no longer survive. These changes have been accompanied by a southwards extension of the Sahara Desert. There are now areas in East Africa which are close to becoming deserts, and it is even possible that Lake Victoria, also a shallow lake, will be greatly reduced in size if irrigation and power generation uses are not controlled. Lakes Rukwa and Manyara are already much smaller than they used to be.

Africa needs to prepare itself for more storms and hurricanes and unseasonal heavy rains, by preserving forests, planting trees and protecting soils which could be threatened by erosion. It also needs to prepare for droughts and long periods without rain. This will require the construction of dams, small and large, to hold back water till it is needed, careful measures to manage the flows in rivers to prevent them drying up, and steps to limit the damage to infrastructure from floods and heavy rain. African cities should not be permitted to expand onto floodplains or in low-lying coastal areas. All these steps are in addition to any commitments African countries make to reducing the quantities of greenhouse gases entering the air, through use of renewable energy sources (e.g. from wind or sun and hydro-power, and resources from plants such as green manure) and more efficient use of fossil fuels. Any discussion of agriculture has to take climate change, and its likely implications, into account.

Droughts and floods are not new to Africa. The short case study which follows shows how much knowledge exists in some typical Tanzanian villages on how to survive periods of drought.

CONCLUSIONS

This chapter has described the principles underlying soil management, and crop and livestock production. How these principles are applied in practice by farmers and livestock keepers in East Africa forms the subject of the next chapter.

Case Study 1: How Villages in North-East Tanzania Can Adapt to Climate Change

This case study is based on work supported by the US National Science Foundation under Grant No. 0921952[8].

Climate change on Mount Kilimanjaro, the nearby Pare Mountains, and the plains below is a reality. It shows up in the rainfall and temperature statistics. Villagers and officials are aware of its impact on their lives: there is less rainfall than before, and less water in streams and rivers. Seasonality and intensity of rainfall is said by farmers to be changing.

From 2010 to 2014 a team of Tanzanians and outsiders worked in eighteen villages in Kilimanjaro Region to find out how farmers, herders and fishers understand this climate change in the broader context of the political, economic, environmental, demographic, technological, administrative, legal and social changes they had experienced over the past 20 to 30 years, and how they can adapt to it.

These villages vary: from settlements high in the mountains grappling with reduced yields of coffee and bananas, to areas close to the Pangani[2] River where pastoral Maasai have learnt to combine their pastoralism with small-scale irrigation.

The key conclusion of this project is that people living locally already have a great store of knowledge relevant to climate change and can co-produce new knowledge in partnership with outside specialists. They are neither 'victims' nor 'villains'. Local knowledge is not just 'traditional' or merely 'indigenous technical knowledge'. Knowledge is also social – about how people work together and interact. For example, women may have concerns and priorities which are not shared by men – for them, knowledge can be a source of power which they are not always be able to use.

Interaction and exchange via periodic small markets link the highlands, the mid-slope areas and the lowlands, and facilitate food security. In addition, other food security practices include land-preparation sensitive to micro-climates which use a wide variety of agro-climatically suitable crops – including rice, millet, sorghum, bananas, cassava and maize. Historically, there have been localised food shortages triggered by drought, floods, insect pests and conflict. However, at no time has the whole area suffered from famine – because of agro-climatic diversity.

---

[2] Above the Nyumba ya Mungu dam, this is called the Ruvu River. The Pangani flows into the Indian Ocean near Tanga.

Forests are huge reservoirs of carbon – some of the forests in northern Tanzania hold more than 400 tonnes per hectare. If these forests are lost, all this carbon will go into the atmosphere and increase the instability and variability of the climate. These forests also supply a great range of foods, medicinal herbs, wood for different purposes, and they protect the springs and water courses.

The state is attempting to "adapt" to climate change with large-scale engineering mega-projects such as dams and reservoirs, or through creation of game reserves or national parks, and by inviting overseas companies to come to farm huge tracts of land with industrial technology. But all of these can block the ability of local farmers and herders to engage in small-scale climate change adaptation based on their local knowledge. Such investments and mega-projects often displace the existing farmers, and once displaced they find it difficult to "adapt".

Uncertainty is corrosive. Great-grandfathers and grandmothers knew and adapted to climate variability. But today's greater climate variability comes on top of uncertainly caused by changes in government administration, laws governing land tenure and access to water, and uncertainty about prices. All this combines to cause anxiety and discourages local innovation and investment. Too much of the discourse about climate change is top-down, focuses on a small number of crops, uses unreliable data, and ignores the knowledge and creativity of small farmers and herders. One of the ways that small farmers survive is by growing many different crops. Compared with monoculture, this reduces the labour demands at peak periods, and also reduces risks. The results can be seen in the great diversity and variety of food products in any local Tanzanian market.

Many rural communities and individuals do not have title deeds or clear rights to the land they farm. They need to know where they stand. They do not understand how REDD (an international funding mechanism to pay for Reduction in Emissions from Deforestation and Land Degradation) will work, nor whether they will benefit from REDD payments. They are worried that a new water law will override village control of land. They are worried that there will be nowhere to graze their cattle in the dry seasons. Women are very concerned that they have to go farther and farther away to find firewood, while outsiders come and harvest wood they had considered as belonging to the community, make charcoal, and sell sand – making no payments to the communities who have depended on these assets for centuries.

In the semi-arid lowlands, pastoralism remains the foundation of livelihoods for most Maasai, but, over the last decade, limitations on local forage have coincided with more frequent and severe droughts. Maasai engagement in horticulture and maize cultivation in one of the villages studied began after the construction of an irrigation canal in 1974 and has continued to increase through subsequent development of irrigation infrastructure.

Small-scale irrigation farming is identified by many – both Maasai and more-recently arrived residents from the highlands – as the single most important adaptive practice undertaken in the recent past to lessen the impact of climate variability. Improved maintenance and expansion of the existing small-scale irrigation infrastructure is central to peoples' aspirations for managing future climate variability. Yet plans to expand irrigation may face resistance in areas where cultivation will block pastoralists' access to grazing near rivers and where the rules about access to newly irrigated parcels are not clear. Adaptation should therefore be understood as a political process for which Tanzania's local government system will bear primary responsibility, despite its meagre resources.

The state could do more, for example by rehabilitating small-scale irrigation schemes, and by ensuring that every irrigation project includes ditches which return unused water to the rivers. District administrators can use their contacts to spread information about innovations and good practice. But government policy will have the most impact if it listens to farmers and herders, and works with them to find the innovations which will best lessen risk and provide resilience, both when there is too much rain and when there is too little.

Further reading on open access on the internet

1. For a presentation similar to this book, see *Environmentally Sound Small Scale Agricultural Projects*. Mohonk Trust,1988. http://pdf.usaid.gov/pdf_docs/PNABC328.pdf

2. Natasha Gilbert "Dirt Poor: The key to tackling hunger in Africa is enriching its soil." *Nature* 29 March 2012. https://www.nature.com/polopoly_fs/1.10311!/menu/main/topColumns/topLeftColumn/pdf/483525a.pdf

3. "The Importance of Humus" *Farmers' Weekly* (Zimbabwe) 2 October 2012 https://www.farmersweekly.co.za/farm-basics/how-to-crop/the-importance-of-humus/

4. Brigitte Nyambo *Agricultural Sector Development Programme: Integrated Pest Management Programme*, Government of Tanzania, Revised version 2009, at http://www.kilimo.go.tz/publications/english%20docs/IPMP%20Plan.pdf

5. Peter Neuenschwander "Control of the Cassava Mealybug in Africa: Lessons from a Biological Control Project", *African Crop Science Journal* Vol.2 No.4 pp.369–383. 1994 http://www.bioline.org.br/request?cs94049

6. *The Quiet Revolution: Push–Pull Technology and the African Farmer*. Occasional Paper, Gatsby Charitable Trust, 2005. http://www.spipm.cgiar.org/c/document_library/get_file?p_l_id=17831&folderId=18530&name=DLFE-94.pdf

7. United Republic of Tanzania, *National Adaptation Programme of Action*, United Nations Environment Programme and Government of Tanzania, 2007 https://unfccc.int/resource/docs/napa/tza01.pdf

8. Tom Smucker, Edna Wangui, Ben Wisner, Pantaleo Munishi, Adolfo Mascarenhas, Charles Bwenge, Gaurav Sinha, Jennifer Olson, Dan Weiner, Eric Lovell and others in the Local Knowledge and Climate Change Adaptation Project (LKCCAP) http://tzclimadapt.ohio.edu/

Topics for essays or exam questions

1. Describe in your own words how plants grow, and why without water they only grow very slowly or not at all.

2. Explain the different ways in which soil can move from place to place. Are these always beneficial for agriculture?

3. Explain why humus is important for plant growth, and how the quantities of humus in a soil can be increased.

4. How can soils used for agriculture be protected from short, heavy storms?

5. Discuss the advantages and disadvantages of different ways of controlling weeds.

6. Explain how "shifting cultivation" allowed soils to recover from the removal of plants, and discuss how rotations of crops can be used to maintain their supply of chemicals if soils are cultivated every year.

7. Explain why some bacteria are very helpful in farming, while others are very damaging.

8. Discuss the "push-pull" method of controlling plant diseases described by the Gatsby Foundation. What are its advantages and disadvantages?

9. On p.7 of her report, Brigitte Nyambo writes: "Pesticides are expensive and good management of their use requires skills and knowledge. Proper use of pesticides can contribute to poverty reduction. However, if misused, they can increase the poverty of end users." Discuss the advantages and disadvantages of using sprays to kill insects, and how the resulting risks can be minimised.

10. Discuss the likely consequences of global warming in Africa, and what African countries can do to prepare for it.

11. The Tanzania Government's National Adaptation Programme to Climate Change (see [7] in the "Further reading" list above) suggests that rising temperatures and changed rainfall patterns will lead to a fall in maize yields of up to 84% in central regions. Yet in recent years, maize production in these places has risen rapidly. If you were a farmer in one of these regions, what changes might you make to your farming practices?

CHAPTER 2

Crops and Livestock:
How People Use the Land

Key themes or concepts discussed in this chapter

- The previous chapter discussed what is needed for plants and animals to grow. This chapter relates this knowledge to how people use the different kinds of land in Africa, and the different opportunities it offers for the cultivation of crops and the raising of livestock.
- It presents a simple classification of types of land, based partly on the quality of the soils, but also on the topography and the climate, particularly the extent and reliability of rainfall. (Issues relating to irrigation, i.e. getting water for agriculture in other ways, are held over to the following chapter.)
- It stresses the need for those who farm the land to have secure rights of tenure, so that if they improve their land they know they will not lose it without fair compensation.
- It compares different ways of making sustainable use of the land, especially through conservation agriculture.
- It presents brief notes on the issues and opportunities raised when considering some of the most important crops.
- It gives special attention to livestock, both as a contribution to agriculture alongside crops (mixed farming), and as a farming activity in its own right (pastoralism and ranching).
- It puts all the above into a context not only of global warming – already discussed briefly in the previous chapter – but also of population growth.

FORMS OF LAND USE

There is no single best use for a plot of land. There are always alternatives, and farmers make choices depending on their experiences of what grows well in the local environment, and on the prices they expect to receive for the products they grow (or, alternatively, how much they will have to pay if they need to buy food to feed their families). They are also influenced by activities and ways of growing crops that have worked well for them in the past; by what other farmers are doing and how successful they have been; by any specific features of the land they cultivate (for example, a patch of land which retains moisture for longer than the surrounding land); and by how far they can afford to risk activities which may not produce a good return in poor years.

Farmers with the best prospects are those whose plots have good soils, with reserves of the chemical nutrients that plants need to grow, and a high organic or humus content which helps the soils retain those nutrients, and allows the soils to retain water.

As considered in the previous chapter, the availability of water is often more important than the quality of the soil itself. When there is sufficient water, the heat of the sun enables plants to grow quickly. But the hot sun also causes water to evaporate, both from the surface of the soil and from the leaves of plants (the technical term for this is *evapotranspiration*).

As a general rule, land with average rainfall of less than 500 millimetres (mm) in a year is described as *arid*, i.e. not suitable for crop agriculture. But this statement must be interpreted with care. Rainfall of 500mm is sufficient for growing a wide range of crops in temperate climates – because temperatures are lower, and hence evaporation is less, and because the rain tends to come in small quantities spread over long periods of the year. In contrast, if 500mm falls in four or five big storms in a short period, most of the rain will simply run off the surface, yet more will evaporate in the hot sun, much of the best soil may be washed away, and few crops will grow.

Some crops can survive with little water, either because their leaves have a waxy cover which reduces evapotranspiration (for example sisal), or because their roots are very deep (for example baobab trees). But without water for long periods, most plants will die.

THE LAND AVAILABLE FOR AGRICULTURE IN AFRICA

In broad terms, the land in sub-Saharan Africa can be divided into:

1. **Tropical rain forests.** These are forests where there is rain or mist for much of the year, encouraged by tall trees which are often very old. Large areas of the Democratic Republic of Congo are tropical forest, also the Kakamega Forest in Kenya, the Mabira Forest in Uganda, and small areas around mountains or hills in other parts of Africa. Many species of plants, including spices and herbs, grow in the warm, damp conditions. But these areas are shrinking fast, as trees are being cut down faster than they can regenerate.

2. **Highland areas.** These are areas above about 1,000 metres above sea level, where 1000mm or more of rain per year is expected, spread through much of the year. Where the soils are derived from volcanos (e.g. around Mounts Kilimanjaro, Rungwe or Hanang in Tanzania, or Mount Kenya) they are extremely fertile, and allow intensive cultivation of a very wide range of crops. These include bananas, coffee, fruits and vegetables, and cereals such as maize or wheat. Elsewhere the soils are not so good. Tea and many other crops can be grown provided that extra nutrients are added through organic manure or artificial fertilisers. In recent years many of these areas have been transformed through the coming of quick-maturing hybrid maize (the Southern Highlands, for example, have become the "grain basket" of Tanzania) and through cultivation of round potatoes (also called Irish potatoes), trees, and vegetables.

3. **River valleys and deltas.** Land that floods regularly is mostly highly fertile, because of the nature of the soil and the minerals which are washed down with the flood water. These *alluvial* soils are used for growing rice, and also for quick-growing fruits and vegetables. In recent years, a major change in Tanzania and Uganda has been the increased use of alluvial soils in valley bottoms for rice growing.

4. **Coastal soils.** Much of the soil along the coast is sandy, with uncertain rainfall, but many of these areas have good access to urban markets. Tree crops such as cashew nuts, coconuts, mangos, citrus and other fruits, and a wide range of vegetables and root crops are grown. Charcoal-making for sale in urban areas is another profitable activity along the coast, together with the mining of sand for the construction industry and craft activities to supply urban markets.

5. **African "savannahs" or "miombo woodlands".** These are large, relatively flat areas, mostly above 1,000 metres above sea level, with an average annual rainfall of between 1000 and 1500 mm. Miombo woodlands cover some 45% of the land area of Tanzania (much of the west and south of the country), most of Zambia and Malawi, and much of Mozambique. The soils here are poor, so trees, mainly acacia, do not grow much above 15 metres tall, which allows light to penetrate and grass to grow. The trees give fertility to the land, and if they are cut down and burnt crops may be grown for a number of years; but then yields fall because key nutrients have been used up or drained deep into the soil. The challenges of moving from this "shifting cultivation" to "continuous cultivation", where most of the land is in permanent use, were discussed in the previous chapter.

6. **Semi-arid lands.** These are areas with an average rainfall of about 500 to 1000 mm per year. The natural vegetation is "scrub" – small trees or bushes – as found around Dodoma or Singida in Tanzania. Continuous cropping is difficult and risky – which makes farmers in these areas reluctant to pay for chemical fertilisers or sprays. However, as already noted, Tanzania has recently had a succession of years with unusually good rainfall, so that while there have been food shortages in some areas the country as a whole has produced enough food for the total population. The higher rainfall has made possible a rapid expansion of maize, sunflower, cassava, sorghum, sesame and tobacco, but with unpredictable rainfall the future for cultivation of these crops in semi-arid lands remains somewhat uncertain.

7. **Arid lands.** These are lands where the annual rainfall is more than 200mm but less than 500mm, concentrated in a short period of the year. In any given year the rains may fail or be limited to a few large storms, so the cultivation of annual crops is risky. If crops are grown, special measures are needed to preserve as much water as possible, and crops that are resistant to drought should be selected. Livestock, which can survive long dry periods, often provide the best use of the land. A good example is the area often described as Maasailand, south of Arusha.

8. **Deserts.** These regions have only rare rainfall - in some years none at all, but on average most deserts get less than 200mm per year. Good soil gets blown away. What remains is largely sand or small particles, with little organic content. It is almost impossible to grow crops in such conditions.

USING AGRICULTURAL LAND SUSTAINABLY

The first essential for ensuring that land is productively used, whether for cultivation or for grazing, is that farmers have some form of secure tenure which gives them an incentive to make long-term improvements and to use the land sustainably. The issue of land tenure is mentioned briefly in Chapter 5 on small farms, and covered in more detail in Chapter 6 on larger farms.

Traditionally, in much of Africa, farm lands were prepared for planting using a hand hoe, with which the soil was turned to bury weeds, and clods were broken up to make a smooth seedbed. The seed was usually broadcast, though sometimes planted in rows or mounds, and loosely covered with soil to protect it from birds and to improve germination. Weeding and harvesting were done by hand. Fertiliser was rarely added: when the soil became depleted the farmer moved to a newly cleared piece of land.

This method of "shifting cultivation" is sustainable when land and labour are plentiful. But the area that can be planted is small: typically a household with two adults using hand hoes can cultivate only about 2 ha per year – scarcely enough to feed a family. Furthermore it is hard work (particularly for women, who provide much of the labour), and often hand cultivation leads to late planting and difficulty in controlling weed growth.

In some parts of Africa – such as areas of Tanzania south of Lake Victoria, where cattle are kept on a large scale – ox-drawn ploughs are used to reduce human labour input and to increase the area planted. In the last twenty to thirty years there has been a rapid increase in the use of oxen to prepare land for planting in other parts of Tanzania (see the first article in the Further Reading at the end of this chapter, by Finn Kjaerby [1]). According to a study carried out in Uganda [2], a family with ox-drawn equipment can typically cultivate 3 to 4 ha per year, and the labour requirement for weeding falls to 35 person-hours per ha, compared to 158 person-hours for hand-weeding a crop where the seeds were broadcast. Where crops are planted in rows, seed drills and inter-row cultivators, which may be pulled by oxen or by tractors, may be used to facilitate weeding.

However, ox cultivation is not possible everywhere: in densely populated areas grazing land is in short supply so oxen compete with crops for land. In some areas loss of cattle through drought and diseases has led to declines in oxen-powered crop production. Many small farmers have returned to hoe cultivation, while others have attempted to switch to tractors.

In the 1960s and 1970s there was a rapid expansion of tractor cultivation in Tanzania, using both 4-wheel tractors (often through tractor hire schemes) and small 2-wheel tractors. These allow much bigger areas to be cultivated (typically 40–50 ha per year with a 4-wheel tractor, 10–15 ha with a 2-wheel machine). They improve the timeliness of planting and weeding, and they greatly reduce drudgery. Thus, where land is available, tractors have the potential to lead to increased production.

However, tractor cultivation also introduces extra costs, and both economic and environmental risks. For example:

- the cost of hiring or buying and using the machinery may not be covered by the increased value of production;
- there may be a shortage of skilled people to operate and maintain the equipment;
- there may be few alternative productive uses for the labour saved by mechanisation;
- there may be insufficient demand for the increased production;
- the household may find itself unable to manage the increased area cultivated (i.e. the labour constraint may simply be shifted from land preparation to weeding or harvesting a crop);
- the soil may be compacted by use of heavy machinery, reducing its ability to absorb rainwater and leaving it vulnerable to water and wind erosion;
- clearing the soil for cultivation may contribute to deforestation, loss of moisture and increased risk of erosion.

A case study of the efficient use of tractor cultivation on a large scale co-operative farm in Tanzania, which illustrates both the benefits and the problems that mechanised agriculture can bring, can be found at the end of Chapter 6.

The use of tractors on poor or marginal soil will always be risky, because the farmers have to pay up front to hire or purchase a tractor, but they may not obtain a good return.

Selection of equipment that is appropriate for land preparation and for later stages in the production cycle must ensure that the overall land use system is *sustainable*. That is to say, production levels must be raised while natural resources are conserved and not depleted or destroyed. This need becomes increasingly important in the context of growing population pressure on land and water resources, and of climate change.

A package of practices has evolved to meet these challenges, known collectively by the title *Conservation Agriculture*. It is critical to the future of Tanzania's land use that these practices are understood and, as far as possible, adopted.

CONSERVATION AGRICULTURE

A simple definition of conservation agriculture is "a package of land use practices aimed at raising productivity while conserving natural resources on a sustainable basis".

Conservation agriculture (CA) comprises three key inter-connected components:

1. Minimum or zero tillage: soil disturbance is minimised and compaction is avoided by planting seed directly into uncultivated soil, thus maintaining the capacity of the soil to absorb moisture, minimising the loss of organic matter and saving enormously on energy and labour costs;

2. Maintenance of permanent ground cover: a permeable layer of vegetation (cover crops or crop residues from the previous season) is retained on the soil surface for as much of the year as possible, protecting the soil from the impact of rain, wind and sun and also providing a source of nutrients for soil organisms;

3. Diversification of cropping systems: a range of crops is planted in rotation, sequence or association. These include nitrogen-fixing legumes and trees planted on both farm and field boundaries, which maintain or improve soil structure and fertility and reduce the spread of disease organisms both above and below the soil surface.

Together, these practices build resilience and contribute to sustainable land use by:

- Protecting the soil surface from erosion, moisture loss and physical compaction;
- Returning organic matter to the soil, thus improving the soil's fertility and its ability to absorb water;
- Allowing precision planting of seeds through the surface mulch, followed by precision application of fertilisers, crop-protection chemicals and irrigation water;
- Facilitating the control of weeds, pests and plant diseases.
- Resulting, together, in increased crop yields while substantially reducing the costs of production, the need for labour and the use of fossil fuels.

For a more detailed but straightforward description of Conservation Agriculture, see article [4] in the Further Reading on the Internet at the end of this chapter.

The benefits of CA were first researched and recognised in South America, particularly in Brazil, Paraguay and Uruguay, where, since the 1970s, CA practices have been adopted on some 66 million hectares i.e. 60% of all cultivated land. North America follows, with 54 million hectares now under CA (24% of the cultivated land). Worldwide, the Food and Agriculture Organization of the United Nations (FAO) estimates that about 11% of all cultivated land (a total of 157m ha) is now farmed under CA systems.[5, Table 2]

Africa and Europe have lagged behind other regions in the adoption of CA (they have respectively 0.9% and 2.8% of cultivated land under CA), but the benefits of the practices have become recognised and many more farmers, as well as governments, researchers and support organisations, are experimenting with the adoption of appropriate CA practices.

There are of course constraints to the rapid adoption of CA practices, particularly by small farmers in countries such as Tanzania:

- Several years may be needed for organic matter to be built up and for soil structure and fertility to be restored by surface mulching;
- If weed infestation is severe, chemical herbicides may have to be used initially, which will be costly for small farmers;
- It is difficult for small farmers to retain all their crop residues on the land, as these form an important part of livestock fodder. (Even if a landowner can access alternative sources of animal fodder, it is difficult to stop neighbouring farmers from grazing their animals on the crop residues.)
- Planting seeds through surface mulch requires the purchase of equipment (jab planters – which may be hand-, ox- or tractor-drawn), though these are much cheaper than tractors or ploughs;
- It may be difficult to overturn a tradition of ploughing land before planting that has been built up over many centuries.

Nevertheless, radical changes to traditional land use practices must be made if the challenges of soil degradation, population pressure and climate change are to be met. Many farmers in Tanzania are already adopting sustainable CA practices: case studies in Arumeru, Karatu and Mbeya Districts are described by Shetto et al [6]. The example of Mountainside Farms in West Kilimanjaro is an interesting example of a large farm using CA methods to grow barley for the brewers SAB Miller [7].

In all these examples, farmers have taken the lead in experimenting with new techniques on a step-by-step basis, adapting them to local social, economic and ecological conditions. However, active support from governments, researchers and aid organisations is needed to ensure that the practices are accepted and spread on a large scale. The notes on key crops which follow should be read with this in mind.

THE ISSUES AND OPPORTUNITIES OF SOME KEY CROPS OF TROPICAL AFRICA

Maize. Maize is easy to cook, can be roasted or turned into flour, and easily transported. But it is a risky crop, depending on rainfall at specific points in the growing period. It has become the most popular crop of sub-Saharan Africa It was not always so: for example it became popular in the Southern Highlands of Tanzania only in the 1950s, and in Rukwa in the 1970s. The highest yields are generally obtained in areas above 1,500 metres with reliable rainfall, using hybrid seeds and fertilisers. (Rasmussen [8] shows how, by 1995, a majority of farmers in the Southern Highlands were using fertiliser to grow maize, in what he described as a green revolution for the Southern Highlands).

Maize is central to Tanzania's food security, and this has led, at many times, to bans on transporting it across national boundaries, and sometimes even across regional or district boundaries. In some years this has resulted in some of the maize, either not being purchased or being sold informally across borders, especially in remote areas like Rukwa.

Hybrid maize seeds are sold by private seed companies, and for good results farmers need to buy new seeds each year. Many consumers prefer the taste and cooking properties of traditional varieties, even though their yields are generally lower. So many farmers reserve an area for traditional varieties for home consumption while also growing hybrid maize for sale.

Hybrid maize is also planted where rainfall is lower, such as in parts of central Tanzania, even though yields there are lower and the risk of failure is greater. If subsidies are removed, many farmers in these areas will not be able to use fertilisers.

A genetically modified variety of maize, which is expected to be more drought-tolerant and disease-resistant than currently used varieties, is now being tested at agricultural research centres in Tanzania and Kenya under a programme called WEMA (Water Efficient Maize for Africa). GM seeds are a contentious issue in Tanzania and elsewhere in the region. The new variety is unlikely to be released for several years, but it is important to understand the potential benefits and risks involved. These are discussed in more detail in Chapter 10.

Sorghum. Sorghum was the most widely grown cereal crop in many parts of Africa before maize was popularised and hybrid seeds became available. Sorghum generally has a lower yield potential than maize, and is considered less palatable as a staple food, but it is much more tolerant of drought and is likely to become more important as a food crop as climate change results in rainfall becoming lower and less predictable. Furthermore, higher-yielding hybrid sorghum varieties are being developed to make it more competitive with maize.

Rice (Paddy). Production of rice in Tanzania stagnated in the early 1980s and then rose sharply, exceeding 800,000 tonnes in 2001, one million tonnes in 2003, two million tonnes in 2010 and nearly three million tonnes in 2015. It is grown by about a third of all Tanzanian farmers, and in towns and cities it is widely consumed especially by higher-income earners.

The expansion of rice production is partly a consequence of population growth, which has led farmers to start using the heavier soils in valley bottoms. According to the National Rice Development Strategy, in 2008, out of a total of 681,000 ha, rice was grown on 464,000 ha of "rainfed lowlands", i.e. valley bottoms that flood after heavy rains, and on 200,000 ha of land that was irrigated. (The issues raised by irrigation are the subject of the next chapter.) The Southern Highlands, which includes areas with the most reliable rainfall, produces about a third of Tanzania's rice (from about a quarter of the land used for rice).

The international organisation Africa Rice has developed a new rice variety, *Nerica* ("New Rice for Africa"), which is high-yielding, responsive to fertiliser, resistant to drought, and tall (which makes it easy to harvest). However, many of the traditional varieties – such as "Kyela rice" from south-western Tanzania – have tastes, aromas and cooking properties which are very special, thus earning premium prices. There should be space for both: varieties like Nerica for their yields, and Kyela rice for its taste.

Wheat. Wheat flour is a very convenient food – for breads, cakes, biscuits, etc. However, to grow well most wheat varieties need a period of cool temperatures and then reliable rains – conditions not easily found in tropical Africa. The implication of this is that wheat should be cultivated in areas where reasonable yields are possible, while imports of wheat should be discouraged and taxed. There is a good argument for following Nigeria's example, where it is a government requirement that "wheat flour" should include 10% cassava flour.

Beans. Beans and other legumes, including cowpeas, chickpeas, soya beans, aduki beans, green grams, black grams, navy beans (used to make baked beans), peas and many others, are valuable as sources of protein. They grow quickly, and most of them support bacteria which fix nitrogen from the air. They can be intercropped with maize and many other crops. Often the leaves can also be eaten. Increased demand for these grain legumes, especially for export to East Asia, is encouraging increased cultivation.

Cassava. Cassava is one of the most versatile natural products in the world. It can be used as human or animal food, processed for industrial uses, and fermented for alcohol production. It is drought tolerant and gives farmers acceptable yields, even without fertiliser. Cassava is Tanzania's second most important food crop after maize, but production has stagnated in the last twenty years. The case study at the end of this chapter discusses the challenges of further developing cultivation of this crop in Tanzania.

Bananas. At least two hundred varieties of bananas and plantains are found in East Africa. They are of different sizes, shapes and colours – green, yellow, red and brown. Some are grown mainly for eating, others for cooking, yet others are best for local beer. Some are more resistant to periods of drought than others. Some are more resistant to diseases and nematodes. Some are easier to transport. Bananas provide shade for other crops – maize, beans, cassava, sweet potatoes, yams, coffee, etc. Their roots are deep and hold the soil in place, and their leaves are good food for animals. Bananas and plantains are the main food crop of Uganda, Rwanda and Burundi and many parts of Tanzania. They grow mainly in high altitude areas, but also close to the sea – in fact anywhere where there is sufficient water. All are thirsty, especially some very high-yielding varieties based on plant material from Asia. (For more detail see [9]).

Banana crops are currently endangered by two serious threats: banana wilt and nematode worms. Both of these get into the soil and attack the banana roots. The wilt is caused by a soil-borne fungus and is easily spread, for example on knives and other cutting tools. The nematode worms move swiftly through the soil and can quickly affect a whole plantation, with devastating results for the farmers. The only remedy is to dig up and burn the infected material and all other banana plants nearby, and not to use the soil for at least six months.

Compared with cereals such as maize, rice and wheat, there has been little research on bananas – perhaps because the plants are not grown from seeds, so it is difficult for commercial seed farms or nurseries to make a profit from selling planting material. But the farmers of Uganda and North West Tanzania are paying a high price for this.

Sweet potatoes. In the words of the website of the International Potato Centre: "Sweet potatoes require fewer inputs and less labour than other staple crops. They tolerate marginal growing conditions, such as dry spells or poor soil. They provide more edible energy per hectare per day than wheat, rice, or cassava." The orange-fleshed varieties also help to protect against vitamin A deficiency in humans. In short, sweet potato is a crop which is nutritious, easy to grow and resistant to drought. *Why do we not hear more about them?*

Round or Irish potatoes. In the past round potatoes were mainly eaten by the European and Asian populations, but they have now become an important part of the African diet. Case Study 13 at the end of Chapter 9 describes how production of round potatoes took off in the Njombe and Mbeya areas of Tanzania. These days they are widely available in the cities.

Oilseeds. Sunflower and sesame (simsim) are increasingly produced, responding to the demand for local sources of cooking oil and for export.

Vegetables. A very wide range of vegetables are grown, contributing to balanced diets. These include root crops such as yams, carrots and onions, brassicas such as cabbage and cauliflower, lettuces, the leaves of many legumes and other plants, and other crops such as tomatoes, peppers, aubergine, okra and many varieties of chilli.

A large horticultural industry has grown up in parts of Kenya and Northern Tanzania based on cultivation of high-value vegetables such as green beans, mangetout, baby corn, peppers and aubergines, as well as fruits and flowers, for export to Europe. These crops are grown under tightly controlled conditions, often in greenhouses or under 'polytunnels'. While they generate valuable income and employment opportunities, they are often in direct competition with small farmers for scarce land and water resources, and may not always serve the best interests of the local population.

Fruits. The most reliable fruits come from trees – mangos, avocados, papaya (pawpaw), coconuts, many kinds of citrus fruits, dates, breadfruit, apples, plums, etc. Other fruits grow on bushes, vines or shrubs – for example, grapes, passion fruit, Cape gooseberries, raspberries and blackberries. Yet other fruits are planted annually – for example watermelons, pineapples and strawberries.

Fruit all add variety – and essential vitamin C – to the diet, so the challenge is to develop their markets.

Nuts. Cashew nuts and groundnuts need no introduction. Macadamia, pecan and bambara nuts are examples of nuts which could be grown more widely in Africa. Cashew trees are drought resistant. The crop is vulnerable to fungal diseases, and to control these farmers need to prune their trees and to spray or dust them with sulphur. However, if they are not sure of being paid promptly for their crop, many farmers will not risk the purchase of chemicals. Processing cashew nuts is difficult because of the shape and hardness of the nut and the need to avoid breaking the kernels. If this is done by hand it is very dangerous for the workers, because of the corrosive – and potentially valuable – liquid released from the shell. Case Study 12 at the end of Chapter 9 shows how Vietnam developed a highly productive cashew production and processing industry.

Spices. In some parts of Tanzania many different spice crops are grown. Cloves require a hot, humid climate for much of the year – as on the islands of Pemba and Zanzibar. Cardamom requires good soil and good rain throughout a long season; it grows best in mountain areas. Ginger requires good soil, warmth and plenty of water. Black pepper grows in warm upland areas. Chillies also like a warm climate, but are tolerant of drought.

Sisal. Sisal has hard, waxy leaves that make it tolerant of drought. The removal of the outer green matter from the leaves ("decortication") requires a lot of water. The washed fibres are used for making string and rope, matting, and as insulation material. Cutting the leaves by hand is hard labour. Sisal plantations traditionally employed migrant workers, paid by the task. Now sisal is increasingly grown by outgrowers – saving the plantation owners the costs of providing housing for their labour force. Issues raised by outgrowers and contract farming are discussed in Chapter 8.

Coffee. Until recently coffee was the main cash crop of high altitude areas. Arabica, from areas where the rain is most reliable, has the best quality, while Robusta is more drought resistant. But coffee has been badly affected by declines in the price, especially after the final failure of the International Coffee Agreement to agree on quotas for each country in 1986. This was partly a consequence of new entrants to the market such as Vietnam. Low prices do not encourage farmers to look after their coffee bushes, or to spray them to control fungal diseases.

Cotton. Young cotton plants need moisture, but once established the plant can tolerate a dry climate. Cotton is subject to more pests and diseases than any other major crop, partly because of its long growing season. Pests and diseases affect the leaves, discolour the bolls, and kill the roots. Rain, especially late in the season, causes the insects and funguses to multiply. But insecticides are expensive and dangerous. Good quality hand-picked cotton gets premium prices on world markets, although contamination and adulteration with water, sand or soil have become major problems in many countries.

Tobacco. Flue-cured tobacco has very precise nutrient requirements if it is to achieve good quality. For this reason it is often grown in poor soils where there are few nutrients, and very specific amounts of fertiliser are added. The leaves are then cured in heated barns, which have to be kept at precise levels of temperature and humidity for several days. Tobacco has become Tanzania's most valuable non-food crop.

Tea. The best quality tea comes from young growing shoots - "two leaves and a bud" are plucked. Tea requires regular amounts of rain or mist, and moderate temperatures. Because of global warming, yields are falling in many areas where the crop has traditionally been grown. Most tea is produced on plantations, but small farmers also grow tea as outgrowers, supplying green leaf to factories which also process leaf from large estates.

Grass and forage crops. Elephant grass, Kikuyu grass, and Brachyaria, and legumes such as lucerne (alfalfa) and Gliricidia, are planted to provide fodder for cattle and other animals as well as for improving the soil. This is discussed in more detail below.

The crops described above are only a few of the many that can be grown in different ecological situations. There are many other plants – fruits, vegetables, forest products, nuts, spices, herbs, flowers and cacti (the last three not discussed above) – that can provide income and add to people's quality of life. There are also medicinal plants, plants that protect against certain insects, and different kinds of edible mushrooms. To consider each one fully would require a separate book. But before moving on, there is one other very important form of land use to consider – livestock production.

ISSUES AND OPPORTUNITIES WITH LIVESTOCK PRODUCTION

Mixed farming

Chickens are kept on most farms in Africa. Many people also keep goats, cattle, sheep, pigs, ducks, rabbits, fish and bees. All these animals contribute to farm incomes, improved nutrition, and the health of the soil through their by-products. Farming where crops are grown and livestock is bred is known as *mixed farming*. Mixed farming is common in Europe and many parts of Africa, where grasses or other fodder crops provide feed for the animals, which, in return, fertilise the soil with their manure. Fodder crops are planted as part of rotations, which allow the land to recover from extensive growing of cereals. This is particularly important as a component of the Conservation Agriculture system described above and in Reference [10] in the Further Reading.

On Mount Kilimanjaro and elsewhere in mountain areas of Tanzania, dairy cows and goats are often kept on the farm in stalls, and are fed with crop residues and grass – often carried from far away. The animals provide milk and calves or kids, as well as manure for the banana and coffee crops. In parts of the country where there are larger herds of cattle, some are trained to pull carts or ploughs. Where insufficient fodder is produced locally, farmers purchase bags of manufactured feed, usually based on maize, but with proteins and minerals added for animal growth.

The use of purchased feeds for livestock is controversial, because the grains that are used to make the feeds could be used to feed people. However, the other benefits that flow from keeping livestock on the farm, as well as the value of a varied diet including meat to both rural and urban consumers, mean that livestock will always play an important part in a mixed farming system.

HIGH YIELDING BREEDS

Exotic breeds of cattle, such as Friesians or Dairy Shorthorns, have higher yield potential than indigenous livestock, but they do not do well when temperatures are too high. When introduced to East Africa they put on weight more slowly, and produce less milk, than would be expected in Europe. They are also more vulnerable to disease.

Cross bred cattle, produced by inseminating local cows with semen from high-yielding European or Asian breeds (Charollais and Sahiwal for example), grow faster and produce more milk and meat than indigenous cattle, but they are still less resistant to diseases and drought so have to be treated with care if they are to realise their full potential.

PASTORALISM OR RANCHING

In many arid and semi-arid areas of Africa, which are too dry for reliable growing of crops, there is sufficient grass to maintain cattle and other animals through most of the year. Over decades, breeds have been selected which can survive for long periods without water – for example Zebu and Ankole cattle in East Africa, and N'Dama cattle in West Africa.

The animals most able to resist drought are camels, which are herded by pastoralists in North Africa as well as parts of Somalia and Kenya.

There is a choice to be made between *pastoralism*, where the animals are moved to where grazing is available, and *ranching* where the animals are kept in fenced off grazing areas. The number of animals kept on ranches has to be carefully controlled to prevent over-grazing: the animals are moved to graze on different areas of the land in rotation. In theory ranching allows more productive use of land than nomadic pastoralism, but in practice it has often proved difficult to obtain a commercial return on the investment required for building fences and water supplies, and it is difficult to prevent intruders coming onto the land.

The advantage of pastoralism is that the farmers take the animals to where grass is growing. The main disadvantage is that there is no restriction on the number of animals that may come into an area, so there is often over-grazing, leaving the soil vulnerable to erosion.

COMPETITION FOR WATER

The key question for anyone keeping animals in the arid and semi-arid tropics is how to give them the best chance of surviving the long, hot, dry seasons. Livestock herders know where there are lakes, ponds, springs, wells or other sources of water, but many of these dry up when there are long periods without rain. In the past the areas around such water points were often reserved for dry season grazing, but with the pressure to expand the area under cultivation much of this land is now used for crop production, and is no longer available for grazing. Disputes between pastoralists and crop farmers over access to water and grazing in the dry season are very common.

Cattle can also contaminate water needed for human use. A case can therefore be made for setting aside some boreholes or drinking sources specifically for cattle. However, there are seldom enough of these water sources, so very large numbers of cattle tend to be attracted to a small number of water points, putting both the soil and the water source at risk.

Extra watering points would reduce the pressure on existing sources, but may not allow more cattle to be kept overall – except where there are no water sources at all and a borehole may permit an extra area to be grazed in the dry season.

LIVESTOCK DISEASES

Cattle and poultry keepers have to deal with many diseases and parasites. Newcastle disease is a virus which can wipe out a whole flock of chickens overnight, unless the birds are individually vaccinated. Rinderpest is a virus which can destroy herds of cattle: it can also be controlled by vaccination. Bovine tuberculosis is caused by a bacterium: it is hard to control but not always fatal. Foot and Mouth Disease is caused by a virus, and can be prevented by vaccination, but if animals are infected there is no alternative but to slaughter them, and as quickly as possible, because the virus is highly infectious.

Cows, goats and sheep attract many insects and parasites – especially ticks which can cause great discomfort, as well as being the cause of tick-borne diseases such as East Coast Fever. These can be controlled by immersing ('dipping') the animals in strong insecticides or by spraying the affected parts. This has to be done regularly as ticks quickly return. Trypanosomiasis, or sleeping sickness, is transferred by the bite of the tsetse fly. It is difficult and expensive to treat. However, the flies need shade to reproduce, so cutting down or burning trees controls their numbers. Increased human settlement in recent years has led to fewer areas infected by tsetse fly.

LIVESTOCK SALES

Animals can be thought of as walking insurance policies and banks. They are an insurance in that livestock may well survive even if crops fail. They are a bank because they are a means of storing wealth, and the "interest" they provide in the form of calves and milk, and, for some pastoralists, blood, far exceeds the interest that could be earned from a conventional bank.

This may lead to a reluctance to sell animals – other than in situations where there is an urgent need for money – and this may lead to overgrazing, especially around water points or river banks. However, there is now plenty of evidence which shows that pastoralists sell cattle when they need money for school fees, transport, and domestic items, and they sell them at the optimum time in terms of their growth (see Raikes [11], p.23ff.).

In short, the farming of cattle, sheep and goats should be actively encouraged and supported. They provide the most productive, and certainly the most reliable, use of arid lands. They are sources of protein and they require relatively little investment. Their meat and hides are an important contribution to the economy. They pull carts and ploughs. They improve the soil and the crops that grow on it. And they make possible varied diets. Chicken, rabbits, fish and bees are also very valuable components of the overall farming system.

GLOBAL WARMING AND POPULATION GROWTH

Global warming and population growth are changing the face of the earth. The issues around global warming were summarised at the end of the previous chapter. The position is expected to get much worse unless all countries start reducing their production of greenhouse gases. That is difficult to achieve quickly, not least because there are powerful interests behind gas, oil and coal production. One country on its own cannot stop global warming. If it continues, or increases, then the costs of flood protection are likely to be huge, while deserts will expand, and there are likely to be many more heavy storms or hurricanes and unseasonal rains. Agriculture will become even more uncertain and risky than it is now.

In contrast, population growth can be influenced by governments of individual countries. The extreme case is China, which now has one of the world's slowest population growth rates. This was achieved by not permitting couples to have more than one child. The policy has recently been relaxed a little, but limiting population growth remains a key part of China's economic strategy.

Tanzania has one of the highest population growth rates in the world. There is also rapid urbanisation, but the total population is growing so fast that the numbers living in the rural areas will continue to increase for many years. On average, women are giving birth to around 6 children and this means that there are close to 3% more people to feed every year. Economic growth has been fast – 6 to 7% in recent years. But this has to be shared with the extra population, and extra infrastructure is needed, such as schools and hospitals. So per capita growth is only 3 to 4% per year, and much of that growth never reaches rural areas. It is not surprising that so many people do not feel better off.

At the present time, it is inconceivable that enough jobs can be created in the formal sector, in either urban or rural areas, to employ all the people who need work. So informal employment in agriculture, craft industries, and services will continue. In this context, rapid population growth is not needed, and becomes a brake on economic growth.

Some people argue that increased population is a good thing. But why? Generally in the country there is no shortage of labour. Improvements in public health mean that most children now survive to become adults – so there is no longer a need to have many children to ensure that parents are looked after when they get old. For politicians, more people means more people to keep satisfied – and that will not be easy if there are shortages of land and a lack of good jobs.

Some years ago, the Danish economist Ester Boserup argued that greater density of population would encourage innovation in agriculture. If there is plenty of land, farmers have no need to innovate. But if there are land shortages, farmers have to adopt continuous cropping, with crop rotations and the use of green manure or inorganic fertilisers to maintain fertility, and they have to spend more time protecting their land from soil erosion. They would do more work, but get increased yields out of limited areas of land.

Idris Kikula, in his important doctoral thesis [12] examined how crops were grown after the people in parts of Iringa Region in Tanzania were compelled to live in villages and therefore had less land that they could cultivate. He searched for innovations, and did not find them. But Finn Kjaerby [1] pointed out that the movement of population into villages had made it impossible to continue shifting cultivation, and that farmers had no alternative but to adopt continuous cropping – which meant adopting innovative practices to keep the land productive.

We can conclude, therefore, that there is some merit in Ester Boserup's thinking, but there is nothing automatic about the results. In the situation facing countries in Africa, it is clear that most countries do not need further increases in their populations.

CONCLUSIONS

Land is precious to those who cultivate it and depend on it. Farmers all over the world are attached to their land. On the whole they do look after it well. When they do not, it is usually caused either by desperation – their businesses have failed and they have little alternative – or, sometimes but rarely, through ignorance. There is plenty of evidence that most farmers know how to look after land.

That does not mean that they cannot be assisted in looking after it better, or in getting more production and income from it. As explained above, in the context of growing population pressure and climate change it is essential that farmers use their land sustainably, increasing production and incomes, while not contributing to land degradation. Part 2 of this book looks at how farmers make the most of the land they have (small areas of land: Chapter 5; larger farms: Chapter 6). In particular, farmers must have easy access to markets (Chapter 7) and finance (Chapter 8). Subsequent chapters address the improvements that can be made through research, technical innovation, extension and training.

Case Study 2: Sweet Potatoes – the Perfect Crop?

Many thanks to the International Potato Centre for giving us permission to use this extract from their website [13].

Sweet potato is the third most important food crop in seven Eastern and Central African countries – outranking cassava and maize. It ranks fourth in importance in six southern African countries and is number 8 in four of those in West Africa.

Cultivating sweet potato requires fewer inputs and less labour than other staple crops. It tolerates marginal growing conditions, such as dry spells or poor soil. Sweet potato provides more edible energy per hectare per day than wheat, rice, or cassava. Its ability to produce better yields in poor conditions with less labour makes it particularly suitable as a crop for households threatened by civil disorder, migration, or diseases such as HIV/AIDS.

An estimated 43 million African children under the age of five are threatened by vitamin A deficiency, a condition causing blindness, disease and premature death. Scientists and extension workers at the International Potato Centre in Peru (CIP) promote the consumption of the orange-fleshed sweet potato (OFSP) varieties known to be high in beta carotene (a precursor to vitamin A) in a food-based approach to combating vitamin A deficiency. The challenge is to breed OFSP varieties that meet consumer preferences and can compete with the traditional white- and yellow-fleshed varieties.

The potential of sweet potato has remained largely untapped in sub-Saharan Africa. Average yields are ten times lower among small-scale farmers than those seen among commercial growers with access to irrigation, fertilisers, and credit. CIP is tackling this "yield gap" between subsistence and commercial producers; providing clean planting material and locally adapted improved varieties, tackling damage caused by sweet potato weevils and emerging pests, and training farmers in new, improved agronomic practices.

Grown mainly in small plots by women, sweet potato is often characterised as a poor person's food crop. However, production in sub-Saharan Africa is expanding faster than any other major crop in the region, and important shifts are taking place in the types of cultivars being grown. The crop's rapid expansion over the past decade is due to a variety of factors, including changes in cropping patterns driven by major disease problems with Africa's cassava and banana crops. Other contributing factors include declining farm size, economic volatility, and growth in commercial production.

Further reading on open access on the internet

1.   Finn Kjaerby "The Development of Agricultural Mechanisation in
     Tanzania", in *Tanzania: Crisis and Struggle for Survival.* Edited by Jannik
     Boesen, Kjell Havnevik, Juhani Koponen, Rie Odgaard. Scandinavian
     Institute of African Studies, 1986, pp.173–190. http://www.diva-portal.
     org/smash/get/diva2:274336/FULLTEXT01.pdfand

2.   David Barton; A. Okuni, F. Agobe, R. Kokoi. "The Impact of Ox-Weeding
     on Labour Use, Labour Costs, and Returns in the Teso Farming System".
     Presented at a workshop in Jinja, Uganda, May 2002. https://assets.
     publishing.service.gov.uk/media/57a08d2340f0b6497400168a/R7401b.
     pdf

3.   ECHO. *Best Practice Note on Conservation Agriculture,* October 2016.
     https://www.echocommunity.org/resources/8b34f554-062f-42d3-9ef4-
     88933083faf7

4.   Amir Kassam, Theodor Friedrich, Rolf Derpsch and Josef Kienzle
     "Overview of the Worldwide Spread of Conservation Agriculture".
     Field Actions Science Reports Vol.8, 2015.
     https://journals.openedition.org/factsreports/3966

5.   Richard Shetto and Marietha Owenya (ed): *Conservation Agriculture as
     practised in Tanzania,* 3 case studies. ACT Network, FAO 2007
     www.fao.org/ag/ca/doc/tanzania_casestudy.pdf

6.   John Y. Simpson and Qing Yang Cheong "Commercial Agricultural
     Production in Tanzania: Mountainside Farms Limited". *International
     Food and Agribusiness Management Review* Volume 17 Special Issue B,
     2014, pp.175–180. http://www.ifama.org/resources/Documents/v17ib/
     Simpson-Cheong.pdf

7.   Torben Rasmussen "The Green Revolution in the Southern Highlands"
     in *Tanzania: Crisis and Struggle for Survival* (see [1] above) pp.191–205.

8.   *East African Highland Bananas – Staple Crop of Poorer Smallholders in
     Africa.* IITA/Irish Aid/NUI Galway Ag4D Research Program blog.
     http://eastafricanhighlandbananas.org/

9.   John McIntire, Daniel Bourzat and Prabhu Pingali *Crop-livestock
     interaction in sub-Saharan Africa.* World Bank Regional and
     Sectoral Stdies, 1992. http://documents.worldbank.org/curated/
     en/505681468768678913/pdf/multi-page.pdf

10. Philip Raikes *Livestock Development and Policy in East Africa.*
     Scandinavian Institute for African Studies, 1981.
     http://www.diva-portal.org/smash/get/diva2:274873/FULLTEXT01.pdf

11   Idris Kikula *Policy Implications on Environment – The Case of
     Villagisation in Tanzania,* Nordic Africa Institute and Dar es Salaam
     University Press, 1997. www.diva-portal.org/../FULLTEXT01.pdf&sa=U
     &ei=yDtSU8jPLaj82wWtuYCwCg..

12. International Potato Centre
     http://cipotato.org/research/sweetpotato-in-africa/

Topics for essays or exam questions

1.   "The availability of water is often more important than the quality of the soil." Discuss this statement with relation to any two types of land in sub-Saharan Africa.

2.   "Definitions of aridity based solely on rainfall quantity are of limited use." Explain this statement with examples from Europe and from Africa, and explain what other factors need to be included to determine whether an area is arid.

3.   Why are trees valuable, not just for the products they produce directly, but also for their contributions to improving the soil and growing other crops?

4.   "Alluvial soils in valley bottoms are very suitable for the cultivation of paddy." Explain this statement, and explain why, in much of Africa, these soils have only recently been put to use in this way.

5.   Discuss the issues that arise for a farmer making a transition from "shifting cultivation" to "permanent cropping".

6.   "Many machines, including tractors and combine harvesters, have been developed to save labour in high-wage economies. But this is much less of an issue in Africa, where labour is cheap and available." Discuss whether there has been too much emphasis on the introduction of tractors, and insufficient emphasis on the potential of oxen, in sub-Saharan Africa.

7.   Under what circumstances is the hoe an efficient farm implement for agricultural production?

8.   Compare the advantages and disadvantages for a small farmer of preparing the land (a) with ox-ploughs or (b) with tractors. Would your answer be the same for all parts of the country?

9.   "Conservation Agriculture systems are the best way of conserving resources and sustaining the capacity of soils to produce food." Discuss this statement in relation to your home environment and describe what you understand to be the key components of a CA system.

10. What needs to be done to increase the production of bananas in Africa?

11. Discuss the advantages of "mixed farming" – combining crops and livestock-rearing on the same farm. How can "fodder crops", grown specifically to provide feed for animals, contribute to improvements in the quality of the soil?

12. "'Traditional' systems of livestock herding, far from being 'primitive', represent complex adjustments to the environment." (Raikes [11], p.1). Explain how this is so.

13. "Animals are a resource which is easily underestimated. They provide the most productive, and certainly the most reliable, land use for arid lands." What is needed now to increase their productivity? Discuss the choices available to pastoralists who have to respond to changing environments and consequent shortages of water and pastures?

CHAPTER 3

Water and Irrigation

Key themes or concepts discussed in this chapter

- Flood, gravity, sprinkler and trickle (or drip) irrigation – the differences between them and the advantages and disadvantages of each.
- The problems of large-scale gravity irrigation projects around the world, including water-logging, salinity, problems caused by intensive use of the soil, maintenance of the infrastructure – but above all the need for effective social systems to share out the available water and maintain the systems.
- The need to conserve water available for irrigation, and to use it well.
- The need to recognise that irrigation is costly, not free, and that farmers who get water from irrigation schemes should use it responsibly – and perhaps pay a charge for it.
- The advantages of small-scale irrigation from rivers and streams.
- The importance of the end-users understanding the demands for water for all other uses – for hydroelectricity, industrial use, drinking water, livestock, as well as for agriculture. The serious downsides if over-extraction of water causes a water source to dry up.

WATER IS LIFE

"Water is life." Without water even the hardiest of plants will eventually shrivel up and die. Water provides the mechanism through which plants absorb nutrients from the soil, and it is a key component of *photosynthesis* – the basic process by which plants create carbohydrates or sugars by taking energy from the sun and carbon dioxide from the atmosphere.

Most plants are *rainfed* – they depend on rain which falls directly on the land. But some are greatly assisted by irrigation – where water from one place is moved to where a crop is planted. In some parts of the world this has a fundamental influence on what is grown and how it is grown.

Almost all the water available for plants comes from rain or snow. Most of this has fallen fairly recently, but some of the water which comes from boreholes and springs fell as rain hundreds, or sometimes thousands, of years ago. It may have travelled a long distance underground – for example much of the water found in oases in deserts. If it is removed today, it may not be replenished for many years. It is not then a renewable resource.

NATURAL OR FLOOD IRRIGATION

This is the simplest form of irrigation. When there is heavy rain, rivers and valley bottoms flood. Farmers plant their crops (or transplant seedlings, e.g. paddy) before the land floods, or while the water is at a suitable depth. The water may then cover the soil for several weeks. After that the crops depend on natural rainfall. The main infrastructure investment, if there is any, is in dykes and embankments to channel the water to where it is needed, and sometimes to keep it away from other places. Floodwater is very fertile because it includes soil particles – which are rich in organic matter and chemicals – which have been washed away upstream. Using floodwater for irrigation was developed on the Nile in Egypt, and on the Tigris and Euphrates rivers in present-day Iraq, more than 8,000 years ago, in China and India a little later, and on the Niger river in West Africa more than 3,000 years ago. This is how Havnevik [2, p.95] describes the present-day agriculture of the Rufiji valley in Tanzania:

> The crucial factor for the fertility of the soil seems to be the nutrients in the silt carried by the flood water and deposited on the top soil. If a field of *mbaragilwa* land [a little away from the river but subject to flooding] has not been flooded for three years, the peasants estimate the yields

to drop to about one half... To take proper advantage of the floods, *mvuli* [flood season] crops must be planted prior to the occurrence of the flood... Planting too early often results in plants wilting because of water shortages. Crops that are planted too late may drown or be washed away by the flood... The dry season [*mlau*] crops grow on previously flooded areas. Planting occurs immediately after the withdrawal of the flood water.

The disadvantage of this kind of irrigation is that it is unpredictable. There may not be sufficient flooding to grow anything. Or there may be so much water that crops get washed away, canals and dykes get destroyed, and so on. A paradox of farming in the Rufiji flood plain is that a big flood will wash away crops that have recently been planted, so that the people are left almost without food. Yet as soon as the floods recede they can plant quick-growing crops such as spinach (*mchicha*), tomatoes, varieties of beans or maize, or watermelons, and within a few weeks they have food to eat – and to sell.

Engineers tend to be suspicious of natural irrigation. They do not like the unpredictability; they want to manage the irrigation system so that they can control the floods. Yet if this is done using dams, then the water settles and drops its soil within the reservoir. The natural fertiliser of the eroded soil ends up at the bottom of the reservoir and not on the fields of farmers. Furthermore the deposited soil fills the reservoir over time, and reduces its capacity to hold water. This has been a problem with the Aswan High Dam on the Nile in Egypt – before the dam was built, the river flooded naturally and fertilised the land, but this stopped when the dam was built. The same will happen if the Rufiji river is dammed at Stiegler's Gorge to produce electricity.

SIMPLE LIFTING DEVICES

Where water flows in rivers which are not far below the level of the surrounding land, then simple pumps may lift it onto the land. The earliest lifting was done by humans. But by 600 BC, pumps on the Nile in Egypt were powered by the force of the water itself. A little later pumps were invented which were powered by animals or windmills. In some places dams were constructed so that the water could be stored before it was needed for irrigation.

These days, pumps are more likely to be powered by small electric or petrol engines, but human and animal power still play a part, especially on small schemes. An improved pump for operation by humans was developed in Bangladesh in the 1980s. This was *the treadle pump* (FAO

[3]).[3] Its power comes from people's legs instead of their arms, and sometimes involves two operators. Soon after, an improved version was developed in Kenya. This pump can be used where the land is not level – as in much of Africa – and it can supply sufficient pressure to power small sprinkler systems. It is portable, so it can be moved away from the well to prevent theft. Such a pump can lift water from a depth of 6 to 7 metres, and can irrigate a small area. Many of these pumps have been distributed or sold in Tanzania.

GRAVITY IRRIGATION

It is often relatively easy to divert water from streams or rivers without pumping. The water is diverted at a place where a canal or channel can be built, and the water flows along the channel more slowly than in the stream or river from which it comes. This water can be released onto fields without the need for pumps. For this reason gravity irrigation is the cheapest form of irrigation to operate – and it makes no contribution to greenhouse gases and global warming.

Gravity has been used for irrigation by groups of small farmers for thousands of years. Large areas of the Andes were irrigated this way by the Incas, and in North America the Hohokam people developed an extensive network of canals and dams well before the coming of Europeans. In Tanzania, the irrigation systems on Mount Kilimanjaro, where canals take water round the mountain to farms far away (see [4]), were constructed long before the colonial invasions. Similar systems were devised in the Pare, Usambara and Uluguru mountains, in many parts of the Southern Highlands, in Kagera, and in many smaller places such as Sonjo, a village north of the Ngorongoro crater which has been studied by archaeologists. Sometimes several farmers are involved, which requires a great deal of *organisation* – to construct the canals, to maintain them every year, and, above all, to control the distribution of the water. If distribution is not well – and fairly – organised, water may be taken by those with farms near the source or spring, leaving little or none for farmers near the end of the canal. Many of the original canals were great engineering achievements. Rock was cut away in places and aqueducts carefully built with stone, sometimes crossing smaller valleys and streams – even in some cases passing through short tunnels. In

---

[3] This report explains the history of treadle pumps since they were invented in Bangladesh in 1981 (p.9), explains how they work (pp.11–15), and what they can do (pp.15–20). It includes a discussion of the Super Money-Maker Treadle Pump which was designed under a project supported by British Aid.

recent years, the use of plastic pipes has made this kind of construction much easier and cheaper.

Gravity schemes can also irrigate on a large scale. In early times, large-scale irrigation was developed by powerful rulers who mobilised the labour of their subjects to construct and maintain large canals. In more recent years such constructions have usually involved investments by governments. Three developing countries have huge areas of large-scale irrigation – China, India and Pakistan. It is not a coincidence that most of this water comes from rivers which are fed from the massive amounts of rain and snow in the Himalaya mountains – the Yangtze and Huang He (Yellow) rivers in China, the Ganges and Brahmaputra in India and the Indus in Pakistan (see Rosen [5][4]). Two other rivers, also fed by the melting snow, run south from the Himalayas: the Irrawaddy makes possible large scale irrigation in Burma; the Mekong runs through Laos, Thailand, Cambodia and Vietnam. Of the rivers in Africa, only the Nile, the Niger, the Congo and perhaps the Zambezi and Limpopo rivers are anything like as big, and none of these are on the scale of the rivers that drain the Himalayan mountains.

GRAVITY IRRIGATION – PROBLEMS

Very large irrigation schemes have other problems. The Yangtze river in China is severely polluted by industries which pour waste into the river. Much of the irrigated land in the Punjab has problems with salination. There are potential conflicts between the users of water. To control flooding, reservoirs need to be as empty as possible when the rains come, so that they can hold back as much of the flood water as possible – but if they are too empty, the fish die, and there will be insufficient water for hydro-electric generation. If too much of the water is used for generation of hydroelectric power, less is available for irrigation. Much of the engineering work is now forty or fifty years old and there are serious problems of maintenance. As Africa develops its irrigation systems, it needs to ensure that any new schemes avoid these problems, and to make sure that existing irrigation and water supply projects are properly maintained.

Gravity irrigation is the cheapest and simplest to manage. Its main disadvantage is that much of the water sinks into the soil and is wasted. If the land is naturally flat, for example the floodplain of a river, it is

---

[4] This paper shows in simple terms the importance of the Himalaya mountains in providing water for the major irrigation schemes in Asia, and the threats and challenges that these rivers face from global warming and other factors.

relatively easy to construct channels that will take the water where it is wanted. Where the land to be irrigated is not flat, then it must be carefully levelled during construction of the scheme. Soil that has been disturbed will resettle in succeeding months and even years. If the levelling is not correct, then when a field is flooded, parts of it may become waterlogged while other parts stay dry. If water lies in a field for any length of time and evaporates due to heat from the sun, then any salts dissolved in it are left on the soil. Over a period the soil becomes salty. This problem of *salinity* has affected irrigation projects all over the world. It was one of the reasons why the ancient civilisations in the Tigris and Euphrates valleys did not survive. It has spoilt large areas of land in India, Pakistan and northern China. In Tanzania it is a problem on the Pangani river. In theory, salination may be dealt with by flooding the land with large quantities of water, and allowing it to drain away and take the salts with it – but sufficient water may not be available, and some land may not drain easily. The problem can be made worse if standing water is used to control weeds, especially if the soils that are irrigated are poorly drained. This is hard for some farmers to accept, because weeding is hard work, and standing water stops the weeds from growing without the need for labour or machinery – but it is wasteful of precious water, and in the long run salination may destroy the land.

There is a danger of taking too much water out of rivers, especially with concrete weirs. Bruce Fox's controversial paper [6] suggests that this was the main reason that the Great Ruaha River in the Tanzanian Southern Highlands started drying up in the dry seasons. This was bad for biodiversity since many plants could not survive, but animals were also finding it harder to get water, and it affected the tourist industry in the Ruaha Game Park. Less water was available for hydroelectric generation downstream, increasing the risk of power cuts in Dar es Salaam and elsewhere. However, Machibya, Lankford and Mahoo [7] showed that the use of water for irrigation was not the only reason for the river running dry in dry seasons, while admitting that this would be less likely if the water available was used more carefully.

Surplus irrigation water should be collected and reused. Thus part of the capital investment in a gravity irrigation scheme should involve the construction of drainage ditches. These collect water that drains off the land and return it to the river from which it came – for use on other irrigated land downstream. It may, however, have heavy concentrations of fertiliser dissolved in it, and if so should not be used for human consumption. The paper by Franks, Cleaver, Maganga and Hall [8] shows how, in the traditional systems of irrigation in the Usangu plains in Tanzania, water drains from one irrigated area to another below – the water is "used" three times.

The biggest challenge facing large gravity-fed irrigation schemes is the creation and sustaining of a fair organisational structure to decide on who gets the water, when they get it, and for how long. Where traditional rulers created and managed irrigation schemes, they also created systems to allocate the water and make sure that plots far from the source got a fair share. If the state is heavily involved, for example in a settlement scheme where land is allocated to farmers who agree to follow agreed rules, then the state will police the rules. If a private company allocates land to outgrowers, it must make sure that the water is properly shared.

Since rules have to be created, with the detail agreed upon, and have to be enforced, it is inevitable that some farmers will not get as much water as they would like. In a year when there is little rain upstream, and therefore less water, everyone should get less. This suggests that distribution should be based on a proportion of what is available at any time, rather than individuals having water rights for absolute amounts. Water must not be diverted unless it is needed. A small irrigation project, involving (say) thirty or forty farmers, needs a simple co-operative arrangement to agree on and police the use of water. But very large schemes need several tiers of administration. In such cases one person controls the gates which regulate the flow into the main canals; different people control gates lower down the system which determine flows into branch canals, and so on, down to small groups of farmers. In each case, those who farm near the source of the water must share the water with those far away. Franks and his colleagues found [8] that those who have the most influence are farmers with substantial holdings of land near the point where water is drawn from the river, as well as some farmers lower down who had built up holdings of land and influence over many years. Those who did not have influence, and whose farms were not close to the source of the water, often had to wait several weeks to get water; those who missed the best times to plant their rice, including a number of women farmers, got much lower yields.

In many traditional schemes in Asia, complex co-operative arrangements developed to review the allocation of water. These were able to discipline farmers who took more water than they were entitled to. They ensured that channels were regularly maintained – for example, removing weeds and silt, and ensuring that weirs and sluice gates were working as they should.

In some schemes, the costs of maintenance are supposedly met by charges for water. But then there has to be agreement on the level of charges (for example should there be higher charges for water that is delivered early in the season?) and payments for water have to be collected. Those who have access to the water need to understand that it is a precious resource, which involves costs and gives those who use it great benefits – so it is reasonable that they should pay something if they use the water.

DAMS

Dams may be built on rivers; the water stored in the resulting reservoirs may be released for irrigation when it is needed. Dams may also be used to generate hydro-electricity and to supply water for human consumption. In some parts of the world, as on the Missouri river in the USA which drains from the snow-covered Rocky Mountains, or the Tennessee Valley which drains from the Appalachian mountains, large numbers of dams have been built. It is good practice to construct dams on the tributaries of a major river before building a large dam on the main river, because then the smaller dams trap the silt and lessen the destructive force of the fast-running water lower down the river. It is also good to ensure that vegetation is cleared before a valley is flooded – since otherwise it may rot without air and thereby contribute to global warming.

The construction of very large dams has become controversial. That is partly because if a dam is built on relatively flat land, the villages and homes of people living along the river will be flooded. It is also because much of the water will be lost by evaporation – especially where the reservoir is large and not very deep. On the other hand, a new dam may create the opportunity for a new fishing industry – though only if sufficient water is kept in the lake throughout the year. (This has been the problem with Nyumba ya Mungu dam in North East Tanzania, where in a number of years the reservoir has dried up, killing the fish.)

BOREHOLES AND GROUNDWATER

If water cannot be obtained from rivers or streams, it may be possible to obtain it from wells or boreholes. There are parts of Bangladesh, India and Pakistan where the groundwater is only a few metres below the surface. It is then not expensive to pump it to the surface and it can be used for irrigation and for drinking – though a major problem in some parts of the world, especially in Bangladesh, is naturally occurring arsenic which makes the groundwater unsafe for drinking. Boreholes are used for irrigation in parts of the USA not close to rivers, for example in California or the mid-West, usually for *supplementary irrigation*, where the main source of water is the rain, but water can be added if needed in periods of drought or when plants are under stress.

But there are problems. When too much groundwater is removed, the land on the surface may sink. If more water is taken out than is needed to replenish the aquifers (underground water sources or rivers), then wells dry up and new ones have to be deeper. Much of the groundwater in Tanzania is already deep below the surface, and expensive to extract, and sometimes difficult to find – as engineers working on water supplies for domestic consumption are well aware.

Where the groundwater is not so deep, boreholes may be sunk without using mechanical drills. The technique is to insert a hollow pipe into the soil, and to drop a heavy weight attached to a cutting tool down the pipe. The weight hits the rock below and cuts into it. The weight is on the end of a wire: it can be pulled back to the surface, and the procedure repeated. Once clean water is found, it may be lifted to the surface using rope pumps: here a rope runs up the inside of the pipe and has washers on it which trap the water and enable it to be lifted. This, however, is very hard work if the water is deep down. This technology is spreading fast because of its cheapness, but it risks wasting water that escapes from the aquifer below the surface if it is not properly used. (For a demonstration video, see [9].)

SPRINKLERS

Where large quantities of water are not available for gravity irrigation, the main alternative is to use sprinklers. These are mechanical devices where nozzles create artificial rain. The water has to come at high pressure, which usually involves pumps. There are two main variations in the technology. "Travelling" systems have nozzles mounted on large gantries which slowly move across the fields (as in Kilombero Plantations, in the Kilombero Valley, part of the Southern Agricultural Growth Corridor of Tanzania). Circular or pivot systems, visible from the air in parts of Zimbabwe and South Africa, have wide booms or beams attached to a central column: the booms, usually driven by electric motors, slowly rotate around the field delivering water to the plants.

This kind of irrigation has the advantage that the land does not have to be levelled, nor do canals or drainage ditches have to be constructed. The precise amount of water delivered can be adjusted to the needs of a particular crop on a particular day. They are still wasteful of water, which is lost by evaporation, though not as wasteful as gravity irrigation. They are not always easy to maintain or to set up, and the pumps are expensive to run and maintain. Temporary systems where the sprinklers are moved in when needed are often used for supplementary irrigation.

DRIP, OR TRICKLE, IRRIGATION

Drip irrigation – also known as trickle irrigation – is the form of irrigation that makes the best use of the available water, but technically it is the most difficult to implement. Very small pipes take the water close to the roots of the plants, where it is released to the plants through tiny holes. In some systems fertilisers may be added to the water.

Trickle (or drip) irrigation was pioneered by engineers in Israel to grow citrus fruit in the Negev desert, where the quantities of water available from the Jordan River and from groundwater were extremely limited (and controversial), and the sandy soil of the desert meant that the water drains through very quickly. This system has enabled large areas of the desert to bloom. It is used in Tanzania, for example on some coffee plantations on the slopes of Kilimanjaro. The system is relatively easy to install for permanent crops such as tea, coffee, or fruit trees such as avocados or oranges, or for sugar cane, but it is difficult to use for annual crops such as paddy. The pipes delivering the water normally stay in the fields permanently. So the crops have to be planted in rows, close to the pipes that cross the field.

Drip irrigation is expensive to instal. It is inflexible compared with other types of irrigation where the water can be diverted or sprayed on different plots. It requires pumps, and the water must be filtered to ensure that the pipes do not get silted up or the small holes blocked. This is difficult on large areas, and in places where the water has a lot of soil etc. in it from erosion on upland slopes. The small holes may also become blocked by roots. Another problem is that rubber parts may decay. A system will need to have a major overhaul, and perhaps be reconstructed, after about five years. This type of irrigation is labour intensive, but it uses water more efficiently than either gravity or sprinkler irrigation. Its greatest potential is for tree crops and where the water is relatively pure, for example borehole water. It will be interesting to see how the method develops in Africa in future years.

OTHER SMALL-SCALE IRRIGATION TECHNOLOGIES

There are other ways to bring water to crops. A simple method which can only be used on a micro-scale is bottle irrigation. Here 2- or 3-litre plastic drink bottles are pierced with a few small holes. They are then filled with water and buried in the ground. The holes are so small that the water takes two or three days to run out, and the bottles can then be refilled by hand. A bottle used in this way can irrigate a circle of about a metre – useful for small quantities of fruits or vegetables near a dwelling house, or a much-loved fruit tree.

Rain water may be "harvested" by the construction of small dams, usually across seasonal water-courses. These have the additional advantage of protecting the land downstream from erosion by fast-running water after a storm. Many such dams have been built in Africa over the years, but often the water has not stayed in them very long. However these days there are improved methods, for example installing plastic linings to stop the water from seeping away – though these may require construction by a specialist contractor.

Rainwater may be harvested from roofs or from impervious rocks. There are some health risks if this water is consumed, especially because of the possibility of contamination from bird droppings, but there are plenty of ways in which such water can be useful – of which small scale irrigation is one.

URBAN AGRICULTURE

Agriculture takes place in the urban areas of Africa, on land close to houses, in gardens, on waste land, and by the sides of roads. Much of this depends on irrigation. Sometimes the urban water supply is used – though the ethics of this are questionable when water is in short supply overall. Or it may come from rain water harvesting, for example from roofs of houses. But often the water comes from local streams, which are very often heavily polluted from human activity. This water is then rich in nutrients, but is also a potential threat to human health, so is safer to use only for the growing of ornamental crops for people's gardens, or trees, than for foods such as *mchicha* (forms of leaves or spinach) for human consumption.

Overall, with increased housing density, the scope for this kind of agriculture is declining. But it remains an important source of supplementary food production. Anyone who can plant a banana or a coconut tree, or use food waste to feed chickens, or maintain a small vegetable garden, will not need to be convinced of its value. Case studies from several African countries are included in the article by Diana Lee-Smith [10].

THE ECONOMICS OF IRRIGATION

Large-scale irrigation is for the long term – it takes many years to recover the costs.

- With gravity schemes, the heavy financial investment is in construction of a weir or other mechanism on the river to divert the water, in building a system of canals, and in levelling the ground. There may also be investment in dams and reservoirs to store water for later use.
- With sprinklers there are the costs of bringing the water to the site, and then the pumps, pipes and sprinkler systems to create the spray and deliver it to where it is needed.
- With drip irrigation, there is the cost of the pipes, pumps, valves and filters, and their maintenance.

Where water is limited, sprinkler irrigation may be preferable to gravity, but the long-term future for irrigation in Tanzania probably lies in drip irrigation.

Irrigation will give the highest returns if the crops grown have high value per unit of weight, or if the available water allows the land to produce two or more crops in a season. But in the dry season the smaller rivers in Africa have only small flows, and water cannot be taken for irrigation without affecting other users downstream. On the Usangu plains in south western Tanzania there is insufficient water in most dry seasons to keep the Great Ruaha River flowing. The case study at the end of this chapter on Dakawa (in the Morogoro region of Tanzania) is of a large scheme where there is seldom sufficient water for irrigation in the dry season.

Irrigation in sub-Saharan Africa is often associated with paddy or sugar cane. But these crops are both heavy users of water. Financial returns may be higher for fruits or vegetables, especially if these can be grown out-of-season.

Agricultural problems are potentially worse on irrigated land. On irrigated plots the plants are close together, and pests, diseases and weeds can spread easily, and may survive, whereas on dry land they would be destroyed by the sun. Two crops a year demand a lot from the soil and hence heavy doses of fertilisers are required. To make matters worse, the irrigation water may wash away inorganic fertilisers, especially on sandy soils. It may also wash away essential trace elements – so these too may need to be added. Thus the risks, in every aspect other than the failure of rains, are greater than for agriculture on non-irrigated land. The risks will be reduced if a variety of crops are grown. A well-run large-scale irrigation scheme needs the back-up of agricultural experts, and a laboratory, to identify technical problems as and when they arise.

But the biggest problems have often been because of the failure of social organisations to enforce the fair sharing of the water and the maintenance of the systems. Doing this is easier if a scheme is developed in association with an existing group of farmers, so that these issues can be discussed from the start. It is more difficult if the infrastructure is already created before farmers are recruited (the "settlement scheme" model), and in such cases it is essential – and difficult – to impose strict conditions. This is hardest on very large schemes.

Many of the large irrigation projects in Africa in the 1970s and 1980s failed for technical reasons – such as miscalculation of the amounts of water available, or failure to level the land properly, leading to salination of the soil. But the more serious failures were with the social organisations, which did not deliver water in a fair way. Much of the investment now is in rehabilitation of old schemes, and they will fail again if the distribution system is not good.

A study of irrigation projects across the "developing world" (Inocencio et al. [11]) analysed data from 314 irrigation projects, comparing the performance of those in Africa – including two in Tanzania – with those in other developing countries. It showed that, on average, the projects in sub-Saharan Africa were more expensive – the construction cost of irrigation was on average over US$8,000/ha (at year 2000 prices) in sub-Saharan Africa but only about US$3,000/ha in the other parts of the developing world. However, this was not the final conclusion of their analysis. They divided the projects into "success" and "failure" projects, with "success projects" defined as those which earned an internal rate of return of 10% or more when they were completed. The breakdown of average construction costs for the two categories – in sub-Saharan Africa and for the rest of the world – are shown in Table 3.1.

Here the figures show investment of US$3,552/ha for "success projects" in sub-Saharan Africa, compared toUS$3,786/ha in other parts of the developing world – almost the same. However many projects in sub-Saharan Africa were failing projects, and these were expensive. The "failure projects" in sub-Saharan Africa cost $17,395 for each hectare, nearly twice the average elsewhere. However, the researchers also found that the percentage of "good" projects was generally increasing over time, and increasing faster in sub-Saharan Africa than elsewhere. They also discovered, as might be expected, that the investment cost per hectare was lower if the total area involved was larger. But they also found that the larger the size of the project, the poorer the performance, in terms of rates of return. So large irrigation schemes are cheaper to construct per unit of area, but, once constructed, they are hard to manage. From this the researchers concluded that the best strategy was to start by designing large projects, but then to divide these into smaller blocks, in each of which the local farmers could be involved in the management of the water.

Table 3.1   Average "hardware" costs in US$ per hectare in 2000, for "Success" and "Failure" irrigation projects

	New Construction	Rehabilitation
Sub-Saharan Africa		
Success Projects	$3,552	$2,303
Failure Projects	$17,395	$9,784
All Projects	$8,188	$5,059
Rest of developing world		
Success Projects	$3,786	$1,415
Failure Projects	$8,924	$3,388
All Projects	$3,183	$1,824

Source: Inocencio et al (2007) Tables 6 and 7

This conclusion is reasonable but it should be treated with caution – it is intrinsically difficult to manage the allocations of water on large irrigation projects. So if there are relatively small schemes with active groups of farmers, and if the projects are technically simple and inexpensive, then they should certainly be implemented. The challenge is to avoid implementing either expensive or technically weak projects, and to ensure wherever possible that farmers are involved in the planning, and that, once the system is up and running, the canals and pumps are properly maintained.

CONCLUSIONS

1. Irrigation will have an important part to play in agriculture in Africa in the future, but it is easy to overestimate the benefits and to underestimate the problems. Engineers are attracted by the technical possibilities of irrigating a particular site, but they often underestimate the technical challenges, especially those of levelling a large area of land. Even worse, they underestimate the social problems, especially getting groups of farmers to work together to share the water that becomes available and to take responsibility for the maintenance of canals, sluice gates, pumps, and other equipment, and to pay a fair price for the water they use.

2. Project designers also often underestimate both the problems and the potentials of better marketing. It is easy to fall back on well-known crops such as rice and sugar, but there may be much higher returns from crops that are more difficult to grow, but use less water, especially if they can be grown outside the main seasons for rain-fed agriculture.

3. Those who plan large irrigation schemes drawing water from rivers should follow plans which look at all the demands for water from those rivers – for drinking water in cities, for animals and people in rural areas, for industrial uses, for the generation of hydro-electricity, and in some cases for fisheries. All of these have legitimate claims on the water. Planners should also take care to avoid rivers drying up because too much water has been drawn off, since this is likely to mean that some plant and animal species will be lost, and this will impact on tourism. They should be aware that in some years rivers have more water in them than in other years. One should never assume that all the water in a river is available for agricultural irrigation.

4. Water is a precious resource, which needs to be conserved in every way possible. This includes reducing evaporation from the soil by covering exposed surfaces with organic matter (mulch), and planting the crops on contours and ridges – as discussed in Chapter 1.

5.   Drainage channels should be part of large irrigation schemes, to return surplus water to the rivers for use downstream, or to send it on to other irrigated fields. Standing water should not be used as a means of controlling weeds – because of the risk of creating salinity, but also because it is a wasteful use of scarce water.

6.   The managers and farmers involved in a large irrigation scheme need to be able to deal with salinity, if this becomes a problem, and also with a lack of the right chemicals in the soil. They need to know what fertilisers are needed, and be able to get hold of them at the right time.

7.   If dams are built, calculations should be made of how long it will take them to silt up, and steps should be taken to conserve soil and prevent it being washed away into rivers. Otherwise the life of dams may be short.

8.   Where there are easily accessible sources of water from streams or springs, small-scale irrigation projects should be developed to make the best use of this water. If these enable crops to be grown outside the main seasons they may be highly profitable.

9.   Gravity irrigation uses water inefficiently; sprinklers are more efficient. But the time has probably come for Africa to follow the semi-arid countries of the Middle East, and adopt drip or trickle irrigation wherever possible.

10. Before any large irrigation project is approved, careful cost-benefit analysis and environmental impact assessments should be carried out. These should include consideration of alternatives, for example means of using less water, or developing the project more slowly. If an irrigation project cannot recover its investment cost in about ten years, then it should not be built, or not built at that time.

Case Study 3: Dakawa Rice Farm

The authors are grateful to Anna Mdee for permission to use this material (see [12]).

In 1981, 2,000 hectares of land about 40 kilometres North of Morogoro in Tanzania was developed by North Korean engineers as a mechanised rice farm. It was divided into 12-acre blocks (just under 5 hectares) irrigated using water pumped from the Wami River.

The Koreans left, and the farm, along with other state farms, passed to the National Agricultural and Food Corporation (NAFCO). Following structural adjustment in the 1980s and privatization in the 1990s, NAFCO got into financial difficulties, and eventually, in 1996, the company went bankrupt. Promises were made that the land would become available to small farmers, including some already farming on nearby swamp-land. But in 1999, by which time most of the land had not been farmed for ten years, it was instead allocated to six large farmers. But their legitimacy was challenged, their titles to the land revoked, and in 2003 the land was passed to UWAWAKUDA, the Ushirika wa Wakulima Wadogo (Society of Small Farmers in Dakawa) with up to 1,000 member farmers. In 2010, as part of the Agricultural Sector Development Programme, USAID, under its Feed the Future programme, agreed on a substantial investment to rehabilitate the scheme. This involved reconditioning the pumps, or replacing them, improving the levelling of the land, and repairing the irrigation channels. In theory no farmer is supposed to farm more than one block, though in practice some farmers have managed to get round this rule.

But there are many problems. In particular, the water in the Wami river in most dry seasons is too little for it to be extracted. That means that, in most years, at the time of year when irrigation is most needed, the pumps cannot be used. It is therefore only possible to get a single crop in a year.

The project was researched by Christopher Mdee in 2013 and further written up by a team led by Anna Mdee (see [12] pp.15–20). The main conclusion, supported by several engineers, is that the lack of water in the dry season means that the project can never be a commercial success, and that, in that sense, the considerable investment in rehabilitating the pumps has been wasted.

When water is abstracted, there are frequent disagreements about its allocation. Those who farm the blocks closest to the pumps get the water they want. Those far away have to wait a long time, often missing the best time to plant their rice, or they do not get any water at all.

Finally, there are financial disputes over the payments to UWAWAKUDA. These include the electricity bills for running the pumps (T.Shs.150,000 per hectare in 2013 but expected to rise substantially) and the costs of a farm manager employed by the cooperative. At the time of the fieldwork, it was clear that the income was not sufficient to cover these costs. These problems became much worse when the government unexpectedly authorised the import of a large quantity of cheap rice, causing the price of rice earned by the farmers to fall.

This project illustrates some of the hazards of large-scale irrigation. It is expensive, uses precious water, and is only able to produce one crop a year. The investment cannot be justified in commercial terms.

It also illustrates how the social aspects of a project need to be worked through and agreed by all affected before an investment takes place. In particular for large-scale irrigation projects, it is essential that there are well-conceived agreements for allocating the available water, and that all concerned have signed contracts that commit them to these arrangements, before the heavy capital investment takes place. There must then be a process for resolving disputes and enforcing the contracts, and penalties for those who do not comply.

Further reading on open access on the internet

1. Historical origins of irrigation. https://en.wikipedia.org/wiki/Irrigation or www.irrigationmuseum.org/exhibit2.aspx
2. Havnevik, Kjell *Tanzania: The Limits to Development* from Above, Nordic Africa Institute and Mkuki na Nyota, 1993. See especially pp.82–97. http://nai.diva-portal.org/smash/get/diva2:277264/FULLTEXT02.pdf
3. *Treadle pumps for Irrigation in Africa*, FAO 2000 ftp://ftp.fao.org/agl/iptrid/kn_syn_01.pdf
4. Goldsmith, Edward and Nicholas Hildyard *The Traditional Irrigation System of the Chagga of Kilimanjaro*. 1984. http://www.edwardgoldsmith.org/1033/the-traditional-irrigation-system-of-the-chagga-of-kilimanjaro/
5. Rosen, Len *Climate Change and Its Impact on our World's Major Rivers – Part 1: The Rivers of Asia*. 2013. http://www.21stcentech.com/climate-change-impact-major-rivers-asia/
6. Fox, Bruce *An Overview of the Usangu Catchment, Ihefu Wetland, & Great Ruaha River Ecosystem Environmental Disaster*, 2004. http://www.tanzaniasafaris.info/ruaha/Ruaha_River_Disaster.pdf
7. Machibya, Magayane; Bruce Lankford, and Henry Mahoo, *Real or imagined water competition? The case of rice irrigation in the Usangu basin and Mtera/Kidatu hydropower, Tanzania*. https://assets.publishing.service.gov.uk/media/57a08d2140f0b6497400167a/R8064Ruaha10-Real_or_Imagined_Competition-paper.pdf
8. Tom Franks, Frances Cleaver, Faustin Maganga, Kurt Hall. *Evolving outcomes of water governance arrangements: Smallholder irrigation on the Usangu plains, Tanzania*. Working Paper 62, Environment, Politics and Development, Department of Geography, King's College London, 1984. http://www.kcl.ac.uk/sspp/departments/geography/research/epd/workingpapers.aspx
9. *How to build your own borehole.* https://www.youtube.com/watch?v =duv6CVWNNLs
10. Lee-Smith, Diana 'Cities feeding people: An update on Urban Agriculture in Equatorial Africa'. *Environment and Urbanisation* Vol.22, No.2 pp.483–499 (2010) http://journals.sagepub.com/doi/pdf/10.1177/0956247810377383
11. Inocencio, Arlene et al. *Costs and performance of irrigation projects: A comparison of sub-Saharan Africa and other developing regions*. Research Report 109, International Water Management Institute, Colombo, Sri Lanka, 2007. http://www3.iwmi.cgiar.org/wp-content/themes/ultimate/pdf/RR109-final.pdf
12. Anna Mdee with Elizabeth Harrison, Chris Mdee, Erast Mdee and Elias Bahati *The Politics of Small-Scale Irrigation in Tanzania: Making Sense of Failed Expectations*. Future Agricultures Working Paper 107, 2014

https://assets.publishing.service.gov.uk/
media/57a089f6e5274a27b200034d/WP107.pdfN.

13. Hatibu, H.F. Mahoo and O.B. Mzirai 'Relevance of Kenyan Irrigation
Experience to Eastern and Southern Africa'. In Herbert Blank, Clifford
Mutero and Hammond Murray-Rust, eds. 2002. *The changing face of
irrigation in Kenya: Opportunities for anticipating change in eastern and
southern Africa*. Colombo, Sri Lanka: International Water Management
Institute pp.303–329. http://www.iwmi.cgiar.org/Publications/Books/
PDF/Changing_face_of_Irrigation.pdf?galog=no

Topics for essays or exam questions

1. Compare the advantages and disadvantages of irrigation based on
    gravity, sprinklers and trickle systems.
2. Too much irrigation water is wasted. What can be done to make
    greater use of the water available in a country such as Tanzania?
3. Discuss the issues raised by charging farmers for irrigation water.
    How might the level of charge be set?
4. Set out the consequences of a river or a dam drying out in a dry
    season. How may this affect farmers, livestock keepers, fishers,
    domestic consumers and the tourist industry?
5. Many irrigation systems have been constructed without first
    identifying the farmers who will use the land. Why should this
    be done, and how is it possible to ensure that they farm the land
    efficiently?
6. What are the main risks facing farmers who adopt irrigation? How
    may these be lessened or overcome?
7. Rainfall is declining in many parts of Tanzania, for example the
    slopes of Mount Kilimanjaro. If this happens, should there be less
    irrigation (because there is less water), or more irrigation (because
    water is needed to keep plants alive), or better designed irrigation?
8. "A field that cannot use rainwater effectively will rarely use irrigation
    water efficiently. In many farming systems, more than 70% of
    the direct rain falling on a crop field is 'lost' as 'non-productive'
    evaporation, or flows into saline sinks". ([13], p.303). Discuss how
    irrigation water can be used more efficiently.

CHAPTER 4

Agricultural Research

Key themes or concepts discussed in this chapter

- "Technology" – and how it is embedded in machines, products, processes, and people
- How new technologies are created – and who gets the benefits.
- The transfer of technology from one country to another, including the risks.
- The different kinds of agricultural technologies and related research specialisms.
- The role of researchers and research institutes, and the different kinds of research they undertake.
- Why the conclusions from trials on research institutes may not be directly applicable on working farms.
- The benefits of plant breeding, and the associated long-term risks and problems.
- Methods of controlling plant diseases, including the biological control of the insects which cause the diseases.
- The disappointing record of agricultural research in Africa.
- The benefits of working closely with farmers.

Much that is proposed to smallholder farmers in Africa is not acceptable to them... It is too costly, or does not suit their farming conditions... The working hypothesis of this book is that one should first look at what farmers themselves are experimenting with, and then use this as a starting point for joint research and development by farmers and scientists.

Africa has a major resource waiting to be tapped, and that is the creativity of its farmers.

— Tewolde Berhan Gebre Egziabher, Ethiopian environmentalist, from his Foreword to *Farmer Innovation in Africa* [14])

WHAT DO WE MEAN BY "TECHNOLOGY"?

A useful definition of technology (from Collins' Dictionary) is "the knowledge and skills available to any human society for industry, art, science, etc."

Technology may be embedded in *a machine*, such as a power tiller, or in *a product*, such as a new drug – how it works, how it is made, what to do if it goes wrong. It may also be embodied in *a process* that can lead to better production of a product such as a crop or a form of livestock – what chemicals to spray to control a certain pest, how often, and in what quantities. Or being aware of the dire consequences of eating a poisonous plant.

A lot of agricultural technology is embedded in *people*. They have skills which they learnt from their parents when they were children, or at college or university, or when working for others. They know that they have to do certain tasks at certain times, or there will be no harvest. That if they are faced by a certain problem, there are various courses of action that they can take. That there are machines, or chemicals, which can help them, but which may be dangerous to human health if not used carefully, and are often expensive or unavailable. They have detailed knowledge about plants that can be used for medicinal purposes. They use sophisticated procedures for selecting seeds for planting in future years – seeds that will be productive, resistant to drought and diseases, and will produce crops that will give them the taste and cooking properties they want, or be suitable for storing and will get good prices if sold. They know a lot about protecting the fertility of the soil. The case study at the end of Chapter 1 shows how one community in North East Tanzania possesses a great deal of relevant technology about how to react to drought and global warming.

THE CREATION AND TRANSFER OF NEW TECHNOLOGIES

Most innovations are adaptations or improvements of existing technologies, or applications of a process or technique that has been used successfully in a different situation. A good example would be a new use of mobile phone technology. Or a decision to include a legume crop in a rotation because it improves the levels of nitrogen in the soil. Or the breeding of improved seeds, as discussed below. In contrast, totally new products or innovations are few – and usually extremely expensive to develop.

New technologies may result from, or be related to, a scientific discovery in a university or research laboratory. But the process of moving from a theoretical possibility to a product that can be purchased in a shop (or on the internet) is usually long and complicated. A scientist or an inventor may make a prototype and demonstrate that it can work. But others will choose the precise specifications, the materials used to manufacture it, the machines that will process those materials; and then the product will need to be tested, certified for safety and perhaps for effectiveness, before it goes onto the market. This is called "development" – the D in "R&D". Much of this R&D is carried out in the laboratories of private companies. Whoever goes down that road has to pay for the research and development well before there is any income from sales, but will hope to recover that investment when the product or process is marketed. There may be disappointment: many new products never get to the stage of being marketed and many others do not cover their costs. If there is an original component in an innovation, its inventor may apply for a patent, which gives them sole rights to sell the product for a number of years. Some products are protected by other means; for example the precise recipe of Coca-Cola is a closely guarded secret.

Most new products originate in countries which already have strong manufacturing sectors (including China), and many technologies are the property of large companies. A good example is the company Monsanto which developed, and patented, seeds which are resistant to the chemical weedkillers that it also manufactures and sells. It is not easy for countries with small manufacturing sectors to break into this process.

The testing of a product, to ensure that it is safe and does what it claims to do, and the regulation of products such as chemicals or road vehicles, forms an important part of the R&D process.

Those who develop new technologies need to sell the resulting products to make an income. Sometimes the new technologies are not much of an improvement on what is already available. Sometimes they are not adapted to African conditions; sometimes the back-up is not available (for example spare parts may be hard to obtain). The owners may lock the users into a long-term relationship with a particular supplier or company. There have been many unsuccessful attempts to transfer technologies to Africa – and some wonderful success stories also.

AGRICULTURE RESEARCH IN AFRICA

Many problems can be avoided if the technology is developed in Africa for African conditions. For many manufactured products this is difficult, but most African countries have agricultural research institutes which have the capacity to develop new technologies, or to report on how best to use existing technologies. These institutes employ the greatest concentrations of research scientists in the country – matched only by the concentrations of researchers on tropical diseases. Thus, for example, Tanzania is divided into seven zones for agricultural research purposes; each has a core institute for crop and livestock research, where there are experts in the main scientific disciplines who can work together in teams to tackle specific problems. There are also smaller sub-stations in the main producing areas. Other agricultural researchers are employed in universities, especially those with agricultural or scientific specialisations.

Research institutes and universities draw on work being done in other countries, through a large number of internationally financed organisations such as those in the CGIAR group[5]).

---

[5] CGIAR (originally the Consultative Group for International Agricultural Research) was created in 1971. Initially it supported four international organisations that undertook the research that underlay what became known as the Green Revolution. At the time of writing it supports 15 research organisations (one of these, IITA, the International Institute of Tropical Agriculture, whose head office is in Nigeria, is, among many other matters, heavily committed to assisting in research on cassava in East Africa). CGIAR members include the World Bank, FAO, UNDP, the European Union and the main Western donors. Much of its finance comes from the Ford Foundation, the Rockefeller Foundation and the Bill and Melinda Gates Foundation.

Agricultural research includes:

1. **Reactive research, or consultancy.** This means responding to, and giving the best possible advice on problems or challenges which are causing concern to farmers in an area, and where the locally based agricultural officers and their extension staff need help to provide the best answers.

2. **Agronomic trials**, to discover the best way to grow particular crops or manage livestock, and then make general recommendations about how these crops or livestock should be cultivated or looked after.

3. **Plant and animal breeding.** The creation of new varieties of plants, or breeds of animals.

4. **Pest and disease control.**

5. **Soil science, soil conservation, mechanisation and irrigation**

6. **Agricultural economics.** Researchers study what farmers do and what returns they get from different activities. Sociologists and anthropologists study how tasks are allocated in rural areas, which crop varieties are planted and why, how harvests can be stored, how food is prepared.

These are considered below:

1. Reactive research

Farming constantly throws up challenges. Perhaps a crop is not growing well, or suffers from some kind of pest or disease attack. Animals are struggling for some reason. Farmers suspect that they have been given bad advice, or sold poor quality seeds or fertilisers.

A research institute is a community of scientists, and between them they can give answers or advice on many problems, sometimes after undertaking simple tests. Researchers are aware of other work in their areas of specialisation; they write articles for journals, books and teaching material. This means that often they can act as consultants, identifying problems and giving advice.

Provided they are specific, enquiries from the field are very valuable to researchers, because the questions make the scientists aware of problems that farmers are facing. Sometimes a challenge may be raised by an extension worker, or by a large farmer. But some farmers may find it daunting to visit a research institute, or the institute may be far away. Research officers may find out about problems when they visit farms, but limited budgets for travel do not make this easy. These issues are revisited in Chapter 9, which shows how useful extension needs an ongoing and regularly updated relationship between farmers, extension workers and research specialists.

2. Agronomic trials

An agronomic trial is the basic research undertaken to discover how best to grow a crop or to use a pasture.

For example, trials can study:

- When best to plant a crop
- How best to prepare the soil for planting
- The best spacing between plants, and whether to plant on ridges or on the flat
- What levels of fertiliser or other chemical inputs to use
- The impact of different kinds of insecticide or fungicide – when and how they should be used.

Very simple trials may be carried out on a school plot, or a village plot, or on the land of a willing farmer. Thus half a field of tomatoes could be planted with fertiliser, and the other half without. If the fertiliser is doing its job, a higher yield of tomatoes will be visible, without any need for careful calculations! Or half the field could be planted at the recommended date, and the other half two weeks later. Or part of a plot of rice could be planted in rows, and the other half randomly. The farmers will see the differences.

On a research institute the principles are the same, but there is more attention to rigour. The plots can be small, but they must be large enough for a calculation of yield to be made. If the yield for a square metre is known, it can be multiplied up to give the estimated yield for a hectare. More than one variable can be researched, but the plots must be repeated to distinguish between the effects of different treatments. Thus a crop could be planted on three different dates, some with no application of fertiliser, or with one of three different levels of application of fertiliser. To include each possible combination would then require 12 plots: four treatments (including one with no fertiliser), each planted on each of the three different dates. If in addition there were 3 kinds of spacing – the usual spacing in the area, closer together than is common, and a wider spacing – then there would have to be 36 plots in total. Each plot would be clearly marked, and the research officer would need to take great care in calculating the yields from each of them. Many other possible variations could be studied. If many variables are researched in the same trial, the number of plots rises very fast. If the differences in yield are small, then statistical tests are used to show whether or not the results are "significant", i.e. whether if repeated the same conclusions would be expected.

Pasture can be studied in the same way at a research institute with facilities for livestock research – for example, to compare what happens if there are a large number of animals on a plot with what happens if there are few, or if different levels of nitrogen fertilisers are applied.

The results of such trials need to be interpreted with care. This is partly because the researchers must keep all other possible variables the same for all the plots. The simplest way is to adopt a "uniform standard". Thus if a trial is not specifically about weeding, all weeds are removed. If it is not about controlling plant diseases or pests, then all insects or plant diseases are controlled beforehand by spraying. Sometimes crops are irrigated during trials to prevent them suffering stress from lack of water. In contrast to this, a real farmer has to cope – simultaneously – with weeds, or plant diseases or pests, or lack of water. So the trial compares what it is set up to compare, but it is not a realistic model of what happens on a farm. Researchers may conclude that with certain methods a high yield is possible, but it does not follow that a real farmer can get that yield.

A further problem is that all soils are different. The soils at a research institute may have been looked after over many years, so they are not typical of conditions on farmers' fields. So any piece of research may be scientifically valid in the research institute site, given the assumptions that the researchers have made, but that does not necessarily mean that the same results can be achieved on farmers' fields. Thus the results of agronomic trials *suggest* what is good practice; but farmers will want to try out the recommendations themselves before deciding whether or not to adopt them permanently.

Wherever possible, in addition to trials in the research station, agronomic trials should also be carried out on farmers' farms, or, even better, as discussed in Chapter 9, working in partnership with farmers or groups of farmers.

3. Plant and animal breeding

"Breeding" means introducing new genetic material with good properties into existing plants or animals, to be carried forward into successive generations. Individual farmers (or livestock owners) do this when they save the best seeds for the next season, or decide which animals to keep; they choose the biggest, or the most resistant to drought, or the ones that produce crops that taste best. Because farmers (or commercial breeders) are using varieties with the best characteristics, and breeding from them, the genetic material is slowly improved.

This is an example of innovation which is driven by farmers. In many cultures it is carried out by women. Anyone who works the land innovates – they try out different ways of doing the tasks, and if they find an improvement they adopt it. Sometimes farmers learn about improvements from each other.

On a research station, the process is much more systematic, but essentially the same. An improved variety may be created by crossing two varieties, for example by pollinating a local variety which is resistant to a plant disease (such as a blight) with pollen from an improved variety which is very responsive to fertiliser. This will produce a mix of results. The best of these are selected, by hand, and planted again. The most successful plants from the resulting harvest are chosen, and their seeds are planted, and so on. This is repeated until a uniform crop with the desired characteristics is achieved.

Varieties are bred to get high yields. They may also be bred to resist plant diseases or pests, to withstand periods without water, to tolerate salt in the water, to include proteins, to deter predators, or to keep for longer after harvesting. Many new varieties of maize are resistant to rust (a fungus that destroys the leaves and other parts of the plant). The case study at the end of this chapter describes some similar research on cassava in Tanzania.

It takes time to develop a new variety of plant which is high yielding, quick growing, resistant to plant diseases or pests, and incorporates other desired traits. It is seldom possible to develop, test and release a new variety in less than five years, and often it takes much longer.

Breeding a new variety of animal can take twenty years or more, because of the time it takes most animals to grow to the point where they can produce offspring (plant breeding to improve varieties of tree crops or bush crops is also slow, for similar reasons). The biggest concentrations of research on livestock are those studying animal diseases and the parasites that cause them – a major part of veterinary science – and animal nutrition the consequences of feeding animals with different diets. Both of these are likely to produce useful results more quickly than creating new breeds.

Many plants are grown from cuttings, not from seeds: a small piece of stem or shoot is placed in fertile soil, and it starts to grow. This is *vegetative propagation*, and it means that all the plants grown from one plant have the same genetic material. The plant is chosen to have very desirable properties. Often the cuttings are grown in a nursery, as is done with tea, cassava and many fruit trees (mangos for example). Where a plant is slow to develop roots, a common technique is to take a cutting from another plant – one that roots more easily – and then attach (or *graft*) to this a stem from the desired plant. The grafted stem then receives nutrients via the roots of the other plant. This is commonly done with fruit trees such as apples or plums.

The advantage of this method is that the farmer knows exactly what to expect from the plants. The disadvantage is that if one plant is subject to a pest or disease, every other plant will be too – there are advantages in having variation between plants. An extreme case of multiplying a single plant is with genetic modification, discussed in Chapter 10. Here a seed with desirable properties is created in a laboratory, and that single seed is multiplied, so all the seeds and plants that come from it are genetically identical – for good or for bad.

MULTIPLYING AND DISTRIBUTING SEEDS

A new seed variety may be registered by a government seed certification agency. The tiny quantities produced in the research station are then multiplied on a "breeder farm", and the resulting seeds are planted in much larger quantities on "seed farms". After that they can be sold, or distributed, to farmers.

The process is speeded up when genetic material is brought in from elsewhere, but not always with good results. Genetic material already in the country is well adapted – to drought, to local plant diseases, and to the cooking and eating habits of the people. There are, for example, more than three hundred varieties of bananas grown in Tanzania, all with different properties. New seeds brought in from other parts of the world, such as rice varieties bred in Asia, or bananas from Malaysia, may not have the properties that are important locally. The risks increase with any form of "vegetative propagation".

For successful plant breeding it is necessary to have access to a "genetic reservoir" of varieties with different properties. In the past, large numbers of varieties were preserved in research stations or seed banks by carefully planting seeds of each variety. The plants that emerged were pollinated by hand, to ensure that they were fertilised by the same variety. The resulting seeds were then harvested, carefully labelled, and stored. This is a major task; it requires attention

to detail and continuity between research workers, just to preserve the
genetic material.

More recently it has become possible to preserve seeds by freezing
them. But in Tanzania, for example, not all research institutes have
reliable refrigerators. There must be good back-up systems, because if
the power fails, and the seeds warm up, they may be lost – unless they
are quickly planted in the old way.

Some of the new seeds are *hybrids*, including most of the seeds that
define the so-called "green revolutions" in South America and Asia.
(This is discussed in depth in Chapter 10.) Hybrid seeds are the result
of crossing two different varieties. That means that if hybrid seeds are
planted, and seeds from the harvest that results are themselves kept
and planted, the results will not be uniform. So farmers are advised
to purchase new seeds each year. This makes hybrids attractive to
commercial seed companies but is costly for farmers.

Many of the new varieties of cereals are "dwarf" or ""semi-dwarf"
varieties, growing on short stems. The advantage of short stems is
that energy is not wasted producing long stems, and that they are less
susceptible to "lodging", or falling over, in wind and heavy rain. These
plants grow fast and take up fertiliser and water more quickly than
most traditional varieties, and so produce high yields. On the other
hand, without fertiliser they do not grow so well, and they may not
have the cooking properties – smell and taste – that local consumers
are used to. Farmers may, therefore, be resistant to dependence on
these new varieties. Moreover, having only a few varieties would
probably mean the loss of the biodiversity and adaptation to local
circumstances that have been developed by African farmers over
many years.

Starting as long ago as the 1970s, US companies, supported by
USAID, persuaded African governments to adopt seed laws which
would give the "owner", i.e. whoever had created and tested a seed,
the right to charge a fee from anyone who purchases it. In these
circumstances, no seeds may be legally sold until they have been
licensed by a government agency. Getting such a licence is expensive
and complicated, and is easier for a big company than for a research
institute or a university. It is almost impossible for a small farmer.
Where laws on this basis are in place, farmers or local traders who
sell seeds which have not been certified, are breaking the law. In some
places, large companies have certified varieties of seed which were
developed by local farmers, or were very similar to local varieties, and

farmers have even been charged to purchase the seeds that they and their fellow farmers developed over many years!

This kind of seed law and certification is not well adapted to deal with vegetative propagation, even though this is associated with some of the greatest dangers. For example farmers may purchase cassava cuttings or banana plants which have fungal diseases on them, and these can quickly spread to nearby plants. Large areas of bananas in Uganda, Burundi and North Western Tanzania have been destroyed in this way.

There is a let-out. Countries can introduce systems of *Quality Declared Seeds* for some local varieties which have not undergone rigorous testing – see the article by Britt Granqvist [1], and the discussion of Quality Declared Seeds in the paper by Coulson and Diyamett [2]. One reason for permitting this is that, for most non-hybrid seeds, a farmer who has purchased them can keep back some of the harvest for planting the following year – so the farmer only has to purchase the seeds once. The quantities sold are not high and the companies which multiply and distribute them can make more profit from sales of hybrids, where new seeds are needed every year. The intention is that this distinction (i.e. designating some seeds as "quality declared") should apply for only a few years, until such time as the variety is certified. But in reality a full system of certification, for many products, is unlikely to be workable.

4. Control of pests and diseases

Entomologists (experts on insects), plant pathologists (experts on the diseases of plants), and veterinary scientists (experts on the diseases of animals) all have parts to play in researching on pests and diseases, and can give useful advice to farmers. But their tasks are not easy. It is claimed that tropical Africa has more plant pests and diseases than anywhere else on the planet, and new diseases appear all the time – including some brought in from outside the country. (An example of this is the "larger grain borer", *Prostephanus truncates*, which arrived in Tanzania in the lorries which carried food aid to the country in 1978 and '79 and are a big risk to food kept in wooden grain stores [3]).

Plant diseases are also caused by viruses, bacteria, fungi and worms (*nematodes*). The main methods of control have already been noted. Thus creating resistant varieties is an important part of plant breeding programmes. Some pests and diseases can be controlled by planting other plants nearby, which create odours which certain insects avoid, and by less intensive cultivation (as noted in Chapters 1 and further discussed in Chapter 5). The most direct method of control is through the use of chemical insecticides. Thus for example, cotton is vulnerable because it takes a long time to grow and is attacked both when it is wet, and when it is very dry. It is therefore sprayed with heavy doses of powerful insecticides – though these are expensive and potentially risky to human health. Cotton is also grown organically, i.e. without synthetic insecticides or fertilisers, with some success [4]. Another example is coffee berry disease – a fungus which can be controlled by spraying with copper sulphate.

5.   Soil science, soil conservation, mechanisation and irrigation

The properties of different soils are studied by soil scientists, who can often use simple tests in the laboratory to explain why crops fail to grow, or grow poorly. They also study the causes of soil erosion and methods of combatting it, and the effects of mechanisation and irrigation on the soils. For example at Camartec (the Centre for Agricultural Mechanization and Rural Technology) in Arusha, Tanzania, the effectiveness, and the durability of mechanical implements are tested. As noted in the previous chapter, irrigated agriculture brings up special research issues – how to apply irrigation water, deal with salinity problems, etc. All the disciplines of agricultural science contribute to research relating to irrigation.

6.   Agricultural economics

The discipline of agricultural economics grew from work with large farms. Economists studied the accounts of a farm, and worked out the expenditure associated with each crop or productive activity carried out: the costs of the expenditure on preparing the land, planting, weeding, agricultural inputs, harvesting, and sales, compared with the income arising from that crop. This led to the concept of the *gross margin* – the profit made from each crop or activity, usually expressed as a gross margin per hectare (or acre) devoted to that activity.

When economists started working with small farmers, they realised that they could not do this kind of study for every farm. So they collected information about costs and returns from representative samples, or groups, of small farmers. In effect they created a model of an *average* or *representative* farm in an area, and worked out its gross margins. Three discoveries were quickly made:

(1) When a farm has access to a limited quantity of labour, there are constraints on how much of any one crop can be grown. But different crops use labour at different times in a growing season, .so planting many types of crop not only minimises the risks, but usually also makes better use of the labour available. The agricultural economists calculated figures for the profits that would result from different combinations of crops planted at different times, and the discipline known as Farming Systems Studies was born[6].

(2) There was logic in the way farmers took decisions. When information was collected about how farmers allocated their labour, it could be seen that this made sense given the labour constraints and the farmers' need to protect themselves from risk. When farmers rejected an innovation, the researchers often discovered that there were good reasons for this.

(3) Many recommendations which derived from research stations were rejected by farmers, because they did not take account of labour constraints at certain times of the year. Or they were rejected because they involved expenses which farmers could not afford or which increased risk, or because they needed to ensure food supplies before planting crops that would be sold for cash.

These matters, and the techniques of Farming Systems Studies, are discussed in depth in the next chapter.

THE INTERESTS OF LARGE COMPANIES IN AGRICULTURAL RESEARCH

The interests of large international companies which want to profit from new technologies in Africa has already been referred to several times. This motivation is most visible in the production and sales of hybrid seeds, which not only need new seeds each year for good results, but also need extensive additions of fertiliser. The company Monsanto managed to add a gene to some of its seeds which enables the plants to survive unharmed by the chemical weedkiller it produces, glyphosate (marketed under the trade name Roundup[TM]) – and then put very strong pressure on governments in Africa and elsewhere to permit genetically modified seeds to be made legal, so that it could sell these weedkillers. The limitations of this, and the wider issues connected to

---

[6] In the early years, this was known as *Farm Management Studies*. But *Farming Systems Studies* is a more accurate description of what it is about, and more commonly used today.

the Green Revolutions which have taken place in South America and Asia and are being promoted in Africa, are discussed in Chapter 10.

CONCLUSIONS

What is Wrong with Agricultural Research in Africa?

A trial conducted in a research institute should be replicable, in the sense that if the same trial is repeated by different researchers, or in a different year, the results should be the same – because everything other than the properties being researched would have been kept the same.

It does not follow, however, that the resulting recommendations are necessarily the best for small farms, or even that small farmers would be able to reproduce the results. For example in a year when the rains are poor, a recommendation to plant a cash crop such as cotton early may not be feasible for small farmers who need to secure their food supplies by planting maize or rice before they can consider planting cotton. Another example is that the quality of the soil on a research plot may be much better than that on farms nearby.

The only way to get over these problems is to try out what is recommended on a number of small farms, with different soils and rainfall, and to study how the farmers respond: Do they accept what is recommended? And if not, why not? And how may their concerns be met? (See, for example, references [5] and [6] for case studies of how this can work in practice). Unfortunately too many researchers have more interest in the scientific accuracy of their results than in whether farmers can adopt them in the field. But the issues are deeper. Good quality agricultural research requires teams of researchers which need to be maintained over long periods of time. This is expensive, and the results, as with other technological innovations – are uncertain. For these reasons it has often proved hard to maintain budgets for agricultural research institutes, and hence continuity in staffing.

The 2012 paper by Coulson and Diyamett [3] starts with ten recommendations. The first two are about resources and staffing. But others are about the need for communities of scientists to work together, and the challenges of making best use of the resources available from international research agencies. One of the recommendations is about the need for value-chain analysis to be undertaken before expensive research is commissioned (this is discussed in Chapter 7). A value-chain analysis of a particular crop (and preferably also studies of the farming system which show how the crop in question relates to the other crops grown in the area) should identify the main constraints that are holding back its increased production.

If there are problems of diseases, pests, crop storage, marketing or processing, farmers are unlikely to increase production unless these are addressed. Thus it may be more important, for example, for researchers and extension workers to promote varieties of bananas which are resistant to wilts (which attack the roots of bananas) than to promote varieties which increase yields – especially if those varieties also require large and reliable quantities of water. Or if a community is considering adopting the small-scale processing of cassava, it needs to consider not just the costs of processing but the likely returns from the processed chips, which may be quite low and not cover the costs of the inputs which would give greater yields. And there is no point in promoting high-yielding varieties of, say, beans or pigeon peas if the markets cannot absorb them (Ellis pointed this out as long ago as 1982 [15]). In other words, agricultural research to assist small farmers is intrinsically inter-disciplinary – so it is unlikely that a single scientific specialism will provide answers that make a real difference.

The most serious failure of research in Africa is its failure, all too often, to understand the problems faced by farmers, especially small farmers, and to work with them to find solutions to those problems. The quotation that started this chapter states that farmers have a great deal of knowledge about soils, plants (both wild and cultivated), ways of lessening risks, and improving human and animal health. As much research as possible should start where the farmers are, and help them to develop the knowledge they have. A slogan for this, coined by the development specialist Robert Chambers, is *Farmer First!* [16]. His conclusions are further discussed in Chapter 9.

Chambers makes the point that any separation of the extension service from the research institutes (as, for example, presently in Tanzania – where research is run by the national government, and extension services by the District Councils) makes it hard for researchers to work with farmers. It is also important to recognise that research is long-term, and that it will not always succeed. But its potential to help farmers improve productivity and live better lives is undeniable. All these matters are taken further in Chapter 9, on agricultural extension.

Case Study 4: Research and Development of Cassava in Tanzania

The authors are grateful to Catherine Njuguna of IITA in Dar es Salaam for help with this case study.

Cassava is one of the most versatile natural products in the world. Both the leaves and the roots can be eaten. The roots can be dried and then ground up to make flour which can be used for animal or human feeds, or converted to starch which is used in food, in the textile industry, as an adhesive, and in cosmetics. It can also be converted into a sweetener for biscuits and soft drinks, fermented into alcohol for use in cooking or as an alcoholic drink, and further distilled to produce ethanol for use as a biofuel.

Cassava is resistant to drought, and can give farmers acceptable yields even with minimum inputs. Most varieties can be stored in the ground until they are needed. By dry weight it is Tanzania's second most important food crop, after bananas.

Production in Tanzania, in terms both of weight (around 7 million metric tonnes) and area (600,000 hectares) has been roughly static since the 1990s. The major challenges include poor agricultural practices, the use of traditional low-yielding varieties, poor soil fertility, drought, shortage of planting material, and attack by pests and diseases.

The cassava mealybug and cassava green mite were big problems in the 1980s and 90s, but their impacts were lessened by introducing natural predators. Currently, two main viral diseases are ravaging the crops: Cassava Mosaic Disease and Cassava Brown Streak Disease. New varieties resistant to Cassava Mosaic Disease were created and released, but they were found to be susceptible to Cassava Brown Streak Disease. Therefore, there is now a concerted effort across East Africa to develop varieties that have resistance to both these diseases. (See references [7]–[12] for more detail.)

Nigeria expanded its production of cassava very rapidly from the mid-1980s. Yields are roughly double those in Tanzania. The Nigerians established large processing plants and passed a law that required millers of wheat and maize flour to include 10% cassava flour. In partnership with the Nigerian government and other actors, the country jumped to become the world's number one cassava producer.

There is potential for substantially greater cassava production in Tanzania, by developing the whole value chain, but this would require a set of linked changes to be effective. So a team has been set up to ensure that all the issues are tackled together. The team includes the International Institute of Tropical Agriculture (IITA), the Ministry of Agriculture and its research stations, and its Roots and Tuber Crops Improvement Programme, and stakeholders in NGOs and the private sector. Several highly relevant projects are now sponsored by the IITA.

Cassava roots are highly perishable and farmers have traditionally processed them by drying them, or by fermenting and then drying them for home use. These operations were undertaken manually, leading to poor quality and low quantity. The IITA and its partners introduced simple equipment for processing cassava, and have trained farmers to process cassava into high quality flour.

Small-scale processing centres spread across the country. Efforts were made to link them with large factories in Dar es Salaam that could use high quality flour in production of some of their biscuits, a cheaper alternative to wheat. However, the small-scale producers were not able to meet the demands of the quantity and quality that the biscuit factories needed.

For full commercialisation to take off, there is a need to move into large-scale cassava processing. Various efforts have gone into this. These include an ambitious 1 billion USD project sponsored by a Chinese private company to process cassava into high quality flour, starch, and animal feed. This is expected to create a demand and a market for cassava, to create jobs and to give farmers living nearby new and reliable sources of income.

A large plant like this will need regular supplies, and these will be most reliably delivered if there are contracts of some kind with small-scale producers. There will be risks of side-selling if farmers can earn more by selling their crop privately. But, without a reliable market, farmers will be reluctant to increase production, because the market can be flooded and prices become very low. It will not be an easy transition to make.

Conclusions

The following are recommended:

1. Continue with efforts to breed improved high-yielding varieties that are resistant to the two most important viruses, as well as meeting end-users' preferences. These include breeding cassava varieties with high starch content that meet industrial needs. Also continue exploring the use of advances in biotechnology such as molecular-marker-assisted breeding to speed up cassava production.

2. Develop a system that will ensure that the varieties coming out of research can reach farmers. This must be self-sustaining, i.e. not dependent on subsidies from IITA or anywhere else. It will be necessary to strengthen the capacity of the Tanzania Seed Agency to certify cassava seeds and of cassava seed entrepreneurs across the country to mass-produce these varieties and supply farmers with affordable and quality disease-free planting materials.

3. Develop more effective technologies to control pests and diseases, including biological control methods for pests, and for disease vectors such as whitefly and papaya mealybug.

4. IITA and Penn State University are developing a manual phone app to help farmers identify cassava diseases. Once this is rolled out, it will help farmers to correctly identify the diseases on their cassava – an important step towards controlling them.

5. Agronomy: farmers need more information on how best to grow the cassava to get maximum yields, including how to control diseases and pests, appropriate spacing, crops that can be intercropped with cassava (such as legumes that improve soil fertility and also provide protein for the family), when to plant and harvest, and how to control weeds. The Africa Cassava Agronomy Initiative is developing "cassava agronomy recommendation tools" to intensify cassava farming, improve root starch quality and reduce the yield gap.

6. Increase extension work to help farmers to increase their yields. The keys to increasing yields are better rootstock, more intensive planting, and using organic fertilizers.

7. Cassava production, planting, weeding and harvesting and some of the processing such as peeling, are labour intensive. So there is a need to introduce appropriate affordable mechanisation such as mechanical planters and peelers, and efficient weed control methods.

8. There is scope for more processing plants, and further downstream processing, in locations that minimise transport costs. These will need to identify specific market outlets – for example for animal feeds (probably the easiest), pharmaceuticals, soft drinks, low-grade distillation for local use as a cooking or lighting fuel, or high grade distillation for use as bio-fuel.

9. The factories will almost certainly need outgrowers to supply them with large volumes of cassava consistently, all year round. These farmers will require large quantities of clean planting materials and

varieties with high starch content from the breeders. They will then need to accept the long-term security of income which comes from selling to factories, and, implicitly, not selling their crop on the open market – even at a price that gives a better (short-term) return. The factories will probably need to employ extension workers to alert farmers to the new markets that will be opened, and to explain to the farmers the actions they need to take in order to benefit from these new markets.

10. NGOs can develop recipes and foods based on cassava, but also using beans, cowpeas, chickpeas, etc. to provide proteins.

11. There is potential for working with the baking industry to substitute high quality cassava flour for wheat flour in some of their products.

Ponsian Sewando has produced a set of value chains for cassava and cassava products in Morogoro District (see [13]). It would be useful to have similar studies for other parts of Africa, and also studies which compare cassava with other crops, such as maize and cotton or cashew nuts. These studies should seek to understand what happens to all these crops in years of poor rainfall, average rainfall, and more than average rainfall.

Further reading on open access on the internet

1. Britt Granqvist, *Is Quality Declared Seed Production an Effective and Sustainable way to Address Seed and Food Security in Africa?* CTA, Wageningen, Netherlands http://knowledge.cta.int/Dossiers/S-T-Policy/ ACP-agricultural-S-T-dialogue/Demanding-Innovation/Feature-articles/ Is-Quality-Declared-Seed-Production-an-effective-and-sustainable-way-to-address-Seed-and-Food-Security-in-Africa

2. Andrew Coulson and Bitrina Diyamett *The Contribution of Agricultural. Research to. Economic. Growth: Policy Implications of a Scoping Study in Tanzania.* International Growth Centre, 2012. https://www.theigc.org/ wp-content/uploads/2014/08/Coulson-and-Diyamett-final-paper.pdf

3. Michael Cross, "Boring into Africa's Grain: A new insect from Central America is suddenly devastating maize stores in Tanzania", *New Scientist* 16 May 1985 http://books.google.co.uk/books?id=qkUvhYfxedoC&pg=P A11&lpg=PA11&dq

4. bioRe supplement, *Ecotextile News*, October 2009, by John Mowbray. https://www.organiccotton.org/oc/../7e849576186f67ae6e5524d71 ec78409.pdf

5. Ninatubu Lema, and Barnabas W. Kapange "Farmers' Organizations and Agricultural Innovation in Tanzania. The Sector Policy for Real Farmer Empowerment". In: B. Winnink, and W. Heemskerk (eds.) *Farmers' Organizations and Agricultural Innovation, Case Studies from Benin, Rwanda and Tanzania.* Bulletin 374. *Development Policy and Practice.* Royal Tropical Institute (KIT), Amsterdam, 2008. www.kit.nl/sed/wp-content/uploads/publications/908_Background.pdf

6. Ninatubu Lema, Chira Schouten and Ted Schrader (eds.) Managing Research for Agricultural Development, *Proceedings of the National Workshop on Client Oriented Research*, Moshi, 2003. www.kit.nl/sed/wp-content/../1393_tinah_Managingresearchforagriculturaldev1.pdf

7. *Starting a cassava revolution in East and Southern Africa*, IITA Blog, April 2017 http://blogs.iita.org/index.php/starting-a-cassava-revolution-in-east-and-southern-africa/

8. *Cassava Diseases in East Central and Southern Africa: A Major Threat to Food Security.* FAO 2010. http://www.fao.org/fileadmin/templates/fcc/ documents/CaCESA_EN.pdf

9. *Protecting cassava from disease? There's an app for that.* Penn State University, 2018. https://theconversation.com/protecting-cassava-from-disease-theres-an-app-for-that-90026

10. *Project brings ray of hope to fight against cassava viruses in Africa*, IITA Brief, April 2017. http://www.iita.org/news-item/project-brings-ray-hope-fight-cassava-viruses-africa/

11. Scientists make breakthrough in identifying first-ever genetic markers associated with resistance to two deadly cassava viral diseases. IITA Press Release, August 2017 http://www.iita.org/news-item/scientists-breakthrough-identifying-genetic-markers-cassava/

12. *Food Security Livelihoods Risk – Destructive Pest Invades*. IITA Press
    Release April 2015. http://www.iita.org/news-item/food-security-
    livelihoods-risk-destructive-pest-invades-tanzania/
13. Ponsian T. Sewando. "Urban Markets-Linked Cassava Value Chain in
    Morogoro Rural District, Tanzania". *Journal of Sustainable Development
    in Africa*, Vol. 14 No. 3. pp.283–300. http://www.jsd-africa.com/
    Jsda/Vol14No3-Summer2012A/PDF/Urban%20Markets-Linked%20
    Cassava%20Value%20Chain.Ponsian%20Sewando.pdf

Further reading not on open access on the internet

14. Chris Reij,and Ann Waters-Bayer (eds.) *Farmer Innovation in Africa: A
    Source of Inspiration for Agriculture*. Earthscan, 2001. The Introduction to
    this is at https://cgspace.cgiar.org/handle/10568/63785
15. Frank Ellis: *Peasant Economics: Farm Households and Agrarian
    Development*. Cambridge University Press, 2nd edition 1993.
16. *Farmer First: Farmer Innovation and Agricultural Research*, edited
    by Robert Chambers, Arnold Pacey and Lori Ann Thrupp. London:
    Intermediate Technology Publications, 1989.

Topics for essays or exam questions

1. Discuss two examples of technologies that have been created in Africa. How were they created? How were they spread? Who benefitted from these technologies?
2. Explain how agronomic trials are set up on a research institute, and why the results may not be directly applicable on a commercial farm.
3. Set out the advantages and disadvantages of producing improved seeds from (a) selection on a farmer's farm and exchange with other farmers and (b) plant breeding on a research institute with access to plant material from abroad.
4. Many of the new varieties of cereals being promoted by international organisations are hybrids based on the dwarf or semi-dwarf varieties developed in Asia. Discuss the advantages and disadvantages of these for (a) producers in Africa and (b) consumers.
5. Discuss the advantages and disadvantages of biological control as a means of controlling plant diseases, using cassava in Tanzania as an example.
6. Explain why an analysis of the relevant value chain should be undertaken ahead of any programme of plant breeding.
7. Make the case for "Quality Declared Seeds".
8. Agricultural economists carry out "adoption studies" to ascertain why some innovations are adopted by farmers while others are rejected. If you were responsible for such a study, what explicit hypotheses would you investigate?
9. The networks of agricultural research institutes are short of resources. How would you make a case to Treasury for granting more resources?
10. Given that temperatures around the world are rising, and large storms are becoming more frequent, what research would you recommend that might help farmers to adapt to these changes?

PART 2

GETTING THE MOST FROM THE LAND
How to Raise Agricultural Productivity

This section of the book includes chapters on small farms, larger farms, then chapters on marketing and credit.
It concludes with an important chapter on the transfer of agricultural technologies, and the means by which farmers can obtain information that will help them to increase their production, including an assessment of extension services.

CHAPTER 5

Small Farms

Key themes or concepts discussed in this chapter:

- Why there is no simple definition of a small farm.
- What motivates small farmers.
- Why farms – including small farms – are businesses.
- The assets that small farmers control – land, family labour, tools and equipment, and local knowledge – and how they minimise risks.
- How the cultivation of many crops, often in the same field, can minimise risks, improve the soil, improve human nutrition, and help to ensure supplies of food throughout the year.
- The challenges faced by small farmers – from nature, from diseases (of human beings as well as of crops and livestock), from uncertain markets, and from some technologies – and how these can be overcome.
- Other relevant activities which impact on the labour available for agriculture – forestry, fishing, trade, activities relating to tourism, small-scale mining and working for someone else – either close to home or far away.
- An action/research method is presented, through which farmers can use information about their farms to calculate the profit which can be made from each crop or combination of crops, and how this profit can be increased. This can help explain why farmers manage their activities the ways that they do, why they choose particular crops or livestock activities, and what changes might improve their lives.
- Some of the false myths about small farmers, and the unfortunate use of words such as "peasant" or "commercial" as terms of abuse.

WHAT IS A SMALL FARM?

There is no simple definition of a "small farm". At the lower end, the available land should be sufficient for a family to grow most of its own food. A family may also sell agricultural products to get the cash it needs for other purposes, or it may have other means of obtaining that cash. The minimum area required will be less where the soils and the rainfall are good. If the family is big it will need more land. At the upper end of the range, the definition has to be equally vague, because different crops need different areas of land and different amounts of labour; but we can say that a large farm uses more labour than is available from a family, and has more land than is needed to grow food for the family and those who work on it.

Consider an example: on Mount Kilimanjaro the soil is good, the rainfall is reliable, and a small family may obtain most of the food it needs from an acre of land (0.4 hectares) by growing bananas intercropped with other produce, and keeping a stall-fed dairy cow and some chickens. They can sell bananas, maize, coffee and vegetables to earn cash. The cow will eat banana leaves, and grass which can be collected from a nearby communally owned forest or field.

However, in a "semi-arid" area near Lake Victoria, with rainfall averaging only a little over 600mm in a year, most families will need 5 acres (2 hectares) to be self-reliant – five times as much as the family on Kilimanjaro – and more if it is a big family. The farmer near Lake Victoria probably also owns cattle, and these can graze on land nearby – or sometimes far away. Some cattle are sold to get cash. Some are trained to pull carts or ploughs.

This chapter does not depend on exact definitions. It describes what motivates a farm family that grows a variety of crops and keeps livestock, for food and for sale.

LAND

Small farmers need land, and they need to know that it will not be taken away from them except in very unusual circumstances – and that in those circumstances they will either be given other similar or better land, or else fair monetary compensation. The rules about land tenure are discussed in more detail in the next chapter.

In years past, if a farmer needed more land, or if a new household was created by a marriage, the people could apply for new land to farm. But this is much less likely to happen today. This is partly because, in many parts of Africa, the land most suited for cultivation of crops has already been allocated, and most of what remains is earmarked for communal uses (schools, churches or mosques, stores, etc.), or has been acquired by incoming investors. Another problem is that much of the land has been cultivated repeatedly for many years, and is now short of some nutrients and can only be made productive if fertiliser or manure is added. It may also be badly eroded. These problems – and some of the challenges caused by the move from shifting cultivation to continuous cultivation – were discussed at the end of Chapter 2.

REDUCING RISKS

The key to understanding farmers, and especially small farmers, is to recognise that, above all, they face the risk of their crops failing, and if this happens they usually have very little to fall back on.

So to lessen these risks:

- they grow a variety of crops – if one is attacked by insects or monkeys or birds they still have the others.
- they plant varieties which are resistant to drought, even though on average these may not produce the highest yields.
- they plant some crops that grow quickly, and others that mature later.
- they plant at different times; for example with maize, if there is little or no rain when the first planting begins to set seed, the crop will fail, but there may still be time for a second planting if rains come later.

As explained in the previous chapter, these actions may not agree with the advice of researchers, who carefully calculate the best time to plant each crop to get the maximum yield (usually as soon as the soil is soft enough to prepare it for cultivation); but *farmers who spread out their planting are choosing to accept a lower average yield in return for a lower risk that they get nothing at all.*

Small farmers usually plant a mixture of crops in a field. This "*multiple cropping*", or "*intercropping*" has many advantages:

- It makes maximum use of small areas of land.
- There is less need for weeding.
- The crop planted first provides shade for the second crop, and lessens evaporation of water in the early stages of growth.
- The quick-growing crop also protects the soil from erosion. If the roots of the two crops are at different levels in the soil, then the two together will make better use of the nutrients and water available.

- The fact that the density of each crop is lower means that plant pests and diseases spread more slowly (unless the pest or disease attacks both crops).
- But above all, if one crop fails the others are still there – and will have more space in which to grow.

In some circumstances, larger-scale farmers also mix crops in a single field. The main reason they do not do this more often is that dealing with mixed crops can be difficult when there is mechanisation – for example to harvest one crop mechanically without destroying the other. Or if they spray, they may have to spray over a wider area – including on crops that do not need it. The fact that small farmers can mix the crops in a field is one of the main reasons why, in many parts of the world, they often achieve higher yields than large farmers.

FARMING IS A BUSINESS

Small farmers make choices, with the land, labour and capital available to them. They make decisions about which crops to plant, where and how much should be planted, at what times, and using which seeds and inputs. They decide how to prepare the land (whether to use hand labour, tractor or oxen), how often to weed, how to deal with pests or diseases, and where to sell. These are commercial decisions, and some farmers are better at making them than others. Their local knowledge is of great value. Thus most farmers in an area grow many of the same crops, and prepare the land, plant, weed and carry out the other agricultural activities at more or less the same times. They inherit knowledge from their parents and relatives, and broadly copy what they see on nearby farms – although a recent worry is that the coming of new seeds and the inputs associated with them means that much of this knowledge is at risk of dying out. Most farmers also innovate: they try out different ways of producing what they need, and learn from experience what works best.

What small farmers do *not* have is a single simple objective: to make as much money as possible over a number of years. If they had that objective, then in some years they would make large amounts of money, in other years nothing or very little. This last option is not acceptable, because in a bad year they would find it hard to survive. So their objective is *to maximise income subject to a series of constraints*:

- They *must* grow sufficient food to feed their families, even if growing other crops would be more profitable. They will try not to have to purchase food at the end of the dry season, when prices are very high.

- They may not have access to sufficient land for maximising their income.
- They may not have the cash to hire extra labour or apply fertilisers even when it would be profitable to do so.

Women do much of the work on small farms – preparing the land, weeding, harvesting, collecting grass to feed stall-fed cattle. This is in addition to looking after young children, cooking, collecting water and firewood, washing clothes, and keeping the home and the area around it tidy. The division of labour differs in different places, but a common arrangement is for women to take the main responsibility for those crops which are primarily for food, and for men to look after the crops grown primarily to earn cash. Where this is so, it is common for women to have no involvement in some of the key commercial decisions and choices, and to depend on men for the cash they need. The advantages of greater equality between the sexes, and how it can be achieved, are considered in Chapter 11.

At times in the year when fewer labourers are in needed in the fields, they can be employed in improving infrastructure or the environment – for example, maintaining walls, fences, buildings, or clearing the beds of streams or watercourses, improving roads and tracks, or creating structures such as terraces to protect soil from erosion. When there is peak demand for labour – for example to weed crops, or to bring in a harvest – the whole family, including children, will take part. (Those with sufficient cash may pay for temporary labour.) In good years, the whole family can share in the success. In poor years, they will share what little there is. It is a way of using all the available labour. The Russian sociologist Alexander Chayanov called this "self-exploitation", because small farmers are exploiting their own labour.

The pressures on small farmers are immense. Firstly there are the challenges created by the economic system. Of these the most important is uncertainty about the prices they will get for their crops, and their powerlessness to influence those prices (discussed in Chapter 7). Other problems arise from the small scale of their operations, which means that they find it hard to access credit, and also inputs and machinery. These challenges, discussed in Chapters 7 and 8, can often be overcome if farmers join together in "producer groups", or co-operatives to increase the scale of their operations, and negotiate agreements with buyers and suppliers of finance and inputs.

The discussion in Chapter 2 set out much that can go wrong: problems from pests, diseases or predators which eat crops; droughts, or excessive rain and floods. In the longer term, farmers have to deal with the declining fertility of the soil, especially when large quantities of organic matter are taken away. They also have to deal with problems which relate to human behaviour, or to bad luck: illnesses, accidents on the farm or in the home, the deaths of adults or children; violence and family breakups. Any of these can make it difficult or even impossible to continue to plant crops, and a family may become destitute. The poorest farmers may lose most of their contact with the rest of society, and take any path to survive. This may include drifting off to towns (though the pressures there are just as great), or coming to depend on alcohol, drugs or crime.

On the positive side, running their own business gives a farm family a degree of independence. If they see an opportunity which has the potential for improvement without undue increase in risk, they can try it out. But they are not obliged to do what someone tells them to do. They can say no when pressured by the government or other outsiders. And they must live with the consequences, good or bad.

If the pressure is on a whole village, the farmers may resist together – sometimes by means which are very subtle and quiet, but protest may on occasion become violent or involve destruction of property. This kind of "peasant resistance" can be frustrating for government officials or extension workers or the agents of an overseas donor. But anthropologists such as James Scott [8], and Michaela von Freyhold [9] have shown that such resistance does not come from nowhere – almost always there are reasons why farmers reject what they are pressed to do. This is further discussed in Chapter 9.

SMALL FARMERS DO MORE THAN FARM

The importance of earnings in rural areas from sources other than agriculture is easily underestimated.

Debbie Fahy Bryceson studied artisanal (small-scale) mining in Africa, which expanded rapidly after about the year 2000, when the price of gold rose to unprecedented heights. On her calculations (see [4]), around 750,000 families, in the broad arc south of Lake Victoria from Musoma to Shinyanga, Geita and Biharamulo, have decided that digging for gold, however dangerous, is more rewarding than growing cotton. Most are co-workers in small groups which share out any proceeds if they find gold. It is not good for the land, which may be left with piles of stones and dangerous holes in the ground. Whether this kind of mining is sustainable in the long term remains to be seen.

In 2008 Bernd Mueller researched the prevalence of wage labour in the West Usambara mountains of Tanzania [10]. This is an upland area of good rainfall, where there are tea plantations, dairy farms, and where small farmers grow vegetables and fruits, mainly for sale in Dar es Salaam. The richest farmers were getting very good incomes. The poorest group had too little land to be self-sufficient and depended on income earned from working for other people, mostly *kibarua* [labouring] work paid for by the day. He concluded that:

> ... any account describing the economics of the West Usambara Mountains purely from an agricultural viewpoint, is bound to miss important facets without which any socio-economic analysis is doomed to be incomplete, particularly for a rigorous understanding of poverty. Even though farming activities provide a common identifier for the vast majority ... this disguises the fact that their livelihoods and productive activities are much more diversified than is often assumed.

Small-scale agriculture remained important for all these groups – with the poorest cultivating just under an acre, with a small amount of irrigation, and undertaking *kibarua* work related to agriculture – but it was by no means their only source of income.

HOW FARMERS MAKE DECISIONS

1. Gross margin analysis

This section considers the research methods used to understand why farmers choose particular crops, and what alternatives they might consider in order to increase their incomes. Both sociologists and economists contribute to this research.

Sociologists or anthropologists set out to understand the roles of different individuals in agriculture, the skills and crafts they practise, the division of labour within the family, the levels of education, and the way knowledge is passed from generation to generation. Their reports describe the crops grown, the techniques used, how tools are obtained, and how communities respond to droughts and other problems.

Agricultural economists use quantitative methods to study the techniques that farmers use. They measure the areas allocated to each crop, the tasks involved in production, the amount of labour required for each task, the use (if any) of purchased inputs, and the income derived from each crop sold. They use this information to make recommendations about what crops farmers should plant and what inputs they should use.

This kind of information can be collected for an individual farm. The discipline of agricultural economics, discussed as a form of research in the previous chapter, began on *large* farms, where the field areas are known, the farmers keep records of the operations carried out each day or week in each field, and of the money spent on each operation. The farmers know the area of each crop cultivated, the costs incurred (including the labour costs associated directly with a particular operation - for example the wages of a tractor driver), and the income eventually derived from sale of that crop. From this they calculate the *gross (profit) margin* for each activity (crop or livestock) on that farm, i.e. the return (income minus costs) for an acre (or hectare) *used for that activity*. Farmers are then advised by the economists to choose the activity or activities where the gross margins are highest.

If the same crop can be grown in different ways, or at different times, or on different types of soil, each of these will have different gross margins. Thus, on a particular farm, maize planted soon after the rains start, with no fertiliser or sprays, has a certain gross margin. If fertiliser is used, the costs are more (including the extra costs for harvesting the larger crop), but the yield and, hence the net income, should be higher. The gross margin using fertiliser will be different from that when fertiliser is not used; it may be higher; but it could be lower if the fertiliser is ineffective or expensive. In a year when there is a crop or market failure, the gross margin may even be negative. If cultivation can be done in different ways, for example using different machines, or tractors or oxen, each of those will give a different gross margin.

Where family members contribute most of the labour, the economists' conclusions may be presented in one of two ways. In the first method, the net income derived from growing an acre of a particular crop is divided by the number of days of family labour needed for cultivating that acre. This calculation gives a *gross margin per area per day of family labour allocated to that activity* (i.e. it measures the return to the family for a day spent on growing that crop). In the second method, a cost for labour is added to the production costs, representing the "opportunity cost" of the family labour (often this figure is the income that could be earned by the family members if they were doing paid casual labour instead of working on the family farm). This gives the gross margin as the net return in money for a unit of land devoted to a specific activity. Either way *"gross margin analysis" lists the crops or activities that the family members undertake in order of profitability, with the implication that the family should undertake the activities with the highest gross margins.*

The agricultural economists soon realised that, for small farms, there is too much variety, and a lack of written records, to make separate calculations for every farm. So they create what they call an *average*, or *representative*, farm. They select a random sample of farms in an area. For each they ask the farmers how many fields they have, and from that they estimate the average size of farms in that area. They ask how many family members are available to work on the farm – and they calculate the average labour availability for the farms in the sample. They take each crop that is grown, identify the tasks involved in growing it (preparing the land, planting, weeding, spraying, harvesting, etc.) and they ask the farmers in the sample how much time they would expect each of these tasks to take for a given area (usually an acre, or half an acre) at each stage in the growing cycle. They ask each farmer how much of that crop was harvested, and how much money would be received if it was sold. By taking averages of all the farms in the sample they work out gross margins for each activity carried out on their representative (or average) farm. Again the implication is that farmers should carry out the activities with the highest gross margins.

However, as explained in the first part of this chapter, there are very good reasons why most small farmers do not grow just one crop. One of the many reasons is that if there is a peak demand for labour on a single crop, the family may not be able to meet the demand, and some of the crop will go to waste. So if, for example, the peak demand will be for harvesting cotton, the family should not plant more cotton than they can realistically expect to harvest. Gross margins implicitly assume that in such a situation the family would hire extra labour. But that may not be possible if the other farmers in the area are equally busy at that time or if the farmer does not have available cash. And so another technique for estimating potential income was developed. Known as "farming systems studies", it explores how labour and land can be *allocated between several crops*, to get the best return.

2. Farming systems studies

Farming systems studies, also known as farm management studies, calculate the costs and returns (i.e. the net income) derived from the combination of crop and livestock production activities that the farmers engage in, using different amounts of labour at different times of the year.

Continuing the illustration of cotton production, the crop is a heavy user of labour, but it is slow growing. It requires labour for preparing the land and planting, then for weeding, and finally for harvesting. This last is the most labour-intensive operation, and it occurs after most other crops are harvested. There is no point in growing cotton that cannot be harvested, so *the area planted should not be more than can be harvested by the labour available in the relevant month*. For most farms, this will mean that not all the land is planted with cotton. In months other than the month for harvesting cotton, there is therefore likely to be unused labour. These people then work on other, quick growing crops, such as maize, cassava, rice or sweet potatoes – which can be harvested before the month when all the labour is needed to harvest the cotton. The gross margins of these crops are lower than for cotton, but they are worth growing because they will give the farm more income overall. This is typical of the situations modelled in a farming systems study.

This kind of study requires data on how much family labour will be needed in each month for growing a unit area of each crop or activity. Clearly a farmer cannot use more land than is available, or more labour in each month than the family can provide. If the area to be planted with a crop is known or assumed, the total labour needed for that crop in each month can be calculated. If several crops are to be grown, the total labour requirement for each month can be calculated – and if (in any month) that exceeds the total labour available, the farmer should in future grow less of one or more of the crops – or find a way of using less labour on one or more of the crops in the critical month(s). (This assumes that the family is not in a position to hire extra labour). In the simple analysis above, by using some of the land for growing cassava, the family gets more income overall, because it does not waste precious time on cotton it cannot harvest.

Continuing the illustration, the gross margin for rice in an area may be higher than for either cotton or cassava. But the main constraint on growing rice is the shortage of suitable valley-bottom land. So the family will plant all the suitable land with rice – and then use whatever labour remains for work on other crops, including possibly both cotton and cassava.

Many farming systems studies were carried out in Africa in the 1960s and 1970s (see Collinson, [5] and [6]). Most of them discovered that the actual choices made by farmers of how much of each crop to plant were close to the predictions of the models. In this way they demonstrated that there was an economic logic in the decisions that the farmers were taking.

From there it was a small step to suggest improvements or innovations that would increase profit without greatly increasing the level of risk. Farming systems data can be used to model what would be expected if a farmer innovates – for example by adding fertiliser. It can also provide an estimate of the loss that will result if there is a delay in the rains, or unexpected rain late in the season. It can be used to calculate the implications of hiring casual labour to help with a short-term peak labour demand, or of hiring a tractor to prepare the land. It will not fully reflect the richness of the range of crops that can be grown on a small farm. But it is much closer to reality than a gross margin analysis of a single crop.

Gross margin analysis is meaningful for a large farm growing predominantly a single crop, where much of the work is mechanised, and labour is not a constraint because the farmer can hire extra labour if it is needed, and where machines or inputs are also not a constraint because the farmer can hire or purchase them if they are needed. It is a logical approach for deciding whether to purchase a new machine, or to plant a different variety of maize on a farm in the mid-West of America. A farming systems approach is much more suitable for a small family farm in Africa, *where the labour is primarily the farm family and the strategy of the family is to achieve certain minimum objectives while avoiding excessive risks, and to do this by allocating the available labour to many different activities.*

Of course it is not imagined that small farmers will use laptops or spreadsheets to decide what to grow. Broadly, they continue to do what they have always done, making small improvements when they can, and trying out innovations on a small scale. If there is unexpected weather, they will adapt their behaviour accordingly. If there is an outbreak of a plant disease, they will devote resources to lessening its impact. The detail of their actual behaviour depends on what happens in a particular year and the opportunities it gives or denies them. But models derived from farming systems studies are extremely valuable, because they demonstrate the logic that underlies the basic decisions that farmers take in order to make the best use of the labour available at any given time throughout the year.

MYTHS, FALSEHOODS AND THE MISLEADING USES OF WORDS

The final part of this chapter looks at some of the myths, or incorrect uses of language, which have been used to make people believe that small farmers are not productive, or are not capable of improving themselves. None of them are supported by research.

The discussion above mentioned one myth: that small farmers are not rational in their actions; that there is no logic in what they do. An in-depth review of the evidence on the behaviour of small farmers, was carried out by Frank Ellis [11], who worked in Tanzania in the 1980s. He showed that small farmers do not maximise absolute profits – i.e. they do not choose the crop or combination of crops and means of growing them that will give them the greatest income over a period of years. But they do maximise profits subject to keeping risks as low as possible. This means, for example, that they are reluctant to take on credit commitments if there is any doubt about their ability to repay the loans, and also that they choose crops whose prices they expect to be high, even if yields may be low.

There is plenty of evidence that small farmers respond to changes in the prices they receive or in the costs of inputs or machines. A good price for a crop almost always leads to increased areas of that crop being planted the following season (and sometimes therefore smaller areas of other crops), while lower prices mean that they will probably plant less of that crop. Some people think that if prices go up farmers will grow less – because they will get the money they need with less work, but empirical evidence does not support this. These facts have clear implications for marketing, the subject of Chapter 7. In particular, if farmers receive little for growing a crop, they will look for different crops or other ways of making an income. If they have to pay a high price to purchase food during the dry season, then the next year they will plant greater areas of food crops, and will store them till they are needed.

Another myth is that there are large areas of land available for irrigation, and in particular for large-scale irrigation. But, as noted in Chapter 3, most of the rivers in Tanzania are relatively small – and with climate change many of them are now drying up in the dry seasons. Water tables are often low, making boreholes expensive, and the levels of many of the large lakes are falling.

It is also simplistic to criticise pastoralists for being interested only in the number of animals they own, and not in selling them. Pastoralists, like other farmers, need money, and they sell animals – where else does most of the meat that people eat come from? Studies show that these farmers sell their animals at the optimum time, before they are too old. It is true that communal grazing land might be more productive if there were fewer cattle, but from the point of view of a cattle-keeper, their wealth is much safer in the form of cattle rather than in the form of money. As discussed in Chapter 2, communal grazing is a cost-effective means of using the arid land, and much of the semi-arid land, in Africa.

The final damaging myth discussed here is that Africa has very large areas of arable land, which are unused and available for cultivation – with the implication that anyone who wants good land for farming can find it, and that large areas can be allocated to big farms without a loss of production from small farms. This is further discussed in Chapter 6; suffice to say here that there are many parts of Africa where there is little or no spare land, and that, where there is land, it is often in places where there are no feeder roads or bridges, so access is very hard. There are different definitions of arable land, many of which take no account of rainfall: they include land where the soil is suitable for growing crops, but the rainfall is insufficient, making them unsuitable for cultivation. Many of the areas in East and Central Africa which are not used for growing crops are *miombo* woodland where soils are poor – and where farmers benefit from the products of the forest. Recent increases in the production of crops have depended on rapidly increasing use of this land. But this cannot go on indefinitely, especially where soils are intrinsically poor. Other areas are traditional grazing areas, where rainfall and soils are not good, and if the land is ploughed there are big risks of soil erosion.

WORDS USED AS WEAPONS

Some of the words used in the discourse on small family farming in Africa can be misleading. They have become used as weapons, to make judgements about farmers. We now discuss three such words – words which have become so confusing that they are best not used.

The first word is Peasant. Here is a dictionary definition (from Google) of the word "peasant":

1. A poor smallholder or agricultural labourer of low social status (chiefly in historical use or with reference to subsistence farming in poorer countries).
2. (*informal*) An ignorant, rude, or unsophisticated person.

The word "peasant" comes from the French *pays* [countryside] and originally just meant someone who lived in the country. In discussions of feudalism, where large areas of land were owned by feudal lords, the word was used to describe serfs who did not own their land and, in return for using it, had to provide free labour to a feudal landowner, and to pay rent or tax, often a proportion of the crops harvested. But the word was also used to describe independent small farmers, and when feudal estates were broken up, as in Russia with the emancipation of the serfs in 1861, when millions of small farms were created. It was in this context that sociologists such as Alexander Chayanov used it as a term of approval – for a farmer who was mostly self-reliant, innovative, and could produce a surplus from a small plot.

In the 1970s this usage was carried over to describe small family farmers in Africa. But in current English usage a "peasant" is someone who is backward, stubborn, and unwilling to innovate – a term of abuse for someone to laugh at, as in the dictionary definition above. So the word has become a weapon – used by those who do not understand small farming and small farmers. The best advice here is to follow sociologists such as Henry Bernstein – who tries not to use the word. Using the term "small family farmer" avoids most of the problems.

Some writers on Africa think that small-scale farming has no future. They write about a "disintegrating peasantry" or the "unravelling" of the Tanzanian peasantry (eg Bryceson [4]). It is true that there are many pressures on those living in rural areas, that increasing numbers cannot survive from the land and have become destitute, and that many – especially young people – have found new lives outside agriculture, especially in urban areas. But the rural population is still increasing and will do so for at least another generation. It is much too early to write off small farmers – especially when there is no realistic prospect of jobs being found for more than a fraction of them in any other sector.

The next word is **Modern**. The objective of improving agriculture in Africa has often been described as introducing "modern farming" (*ukulima wa kisasa*). It is true that there are many productivity increases that are in theory possible, and that many of these depend on Western science. But it is also true that many innovations have been developed without any input from Europeans. Examples are the irrigation channels around mountains such as Kilimanjaro, the use of green manure, and the cultivation of cotton and its spinning to make cloth. And, as discussed above, there are many good reasons why farmers may not implement innovations proposed by outsiders. Any use of the word "modern" to imply that traditional technologies are inefficient is unhelpful.

The last of the words is **Commercial**. This (or its opposite, non-commercial) is another word used as a term of abuse. Small farmers are told that they "have to be commercial". The implication is that they are not. But, as argued above, small farmers need some cash income as well as food and so they are in fact "commercial". Most small farmers grow crops for sale – though they may have decided that it is too risky to use purchased inputs or to pay for tractors. That does not mean that they are not commercial – it may be exactly the opposite: that they have carefully considered the risks and decided that taking on the responsibilities of credit is too great a risk to accept. One of the challenges for policy makers is to create policies which will reassure farmers that if they invest in inputs or machines they will have a strong chance of getting their money back and more. In that context, it is very misleading to imply that small farmers are not commercial.

The word **Transformation** is also misused, especially if it is used to suggest that small farmers need to adopt what they are advised to do by outside experts. Anyone who thinks that they can make farmers change without first relating to them and understanding their constraints is likely to be disappointed and disillusioned when the farmers do not respond as expected. Once farmers have lost trust in what they are told to do, any form of extension work becomes very difficult. This is discussed at length in Chapter 9. Instead, we should be more positive. Small farmers have a great deal of knowledge. They have aspirations to improve their lives. They see pictures on advertising boards or on television of how others live. They do not want their children sleeping under wet blankets when the roof leaks, or expelled from school because they have not paid fees or charges. They know that they must produce food for their families, and sell products to earn cash. They also know that they have the potential to produce much more. Policy makers must work together with the farmers to allow this to happen.

CONCLUSIONS

This chapter summarises what we know about the motivations and decision-making of small farmers. It shows that what they do is governed by two over-riding considerations: to secure their food supply for the forthcoming dry season (because if they have to purchase food it will cost them a lot of money); and to minimise the risks of failure (because if for any reason things go wrong, they have little to fall back on). So their strategy involves protecting their food supply, and minimising their risks – by planting a variety of crops, at different times, and then looking for good ways of earning money to pay for what they cannot produce themselves, all the time being very cautious when they take on debt.

Within this framework, small farmers aim to maximise their profits, planting more of a crop whose price has risen, and allocating the labour available to a combination of crops which they expect to give them the best returns. They also need reliable and efficient marketing arrangements for their crops (see Chapter 7), fair arrangements for purchases of machinery and chemicals (Chapter 8), and information about possible innovations (Chapter 9).

An action research and training approach, developed by Frédéric Kilcher, is summarised in the case study below.

This chapter has shown that, if the conditions are right, small farmers can increase their production and produce surpluses. In the context of declining world prices for many agricultural products, to achieve long-term growth in agricultural production at rates higher than the growth rate of the population is an achievement. Governments in Africa that achieve this should celebrate the successes of their small farmers.

But for this to be recognised, policy makers and extension workers need to respect small farmers, and to avoid language or words that are misleading, and especially to avoid any kind of language that implies that small farmers cannot improve themselves.

Case Study 5: Profitability Analysis of Small Farms in Northern Tanzania

We are deeply indebted to Frédéric Kilcher (CIO of AGRIinsight) who drew our attention to the work described in a report [7], by the Grow Africa Smallholder Working Group.

Farming as a Business training sessions normally involve three intensive 6-hour sessions with groups of farmers.

The farmers work individually or in small groups to create quantitative figures about the crops they have grown recently, the costs involved, the labour used in different months of the year, and the resulting profit, or gross margin, for each crop (or combination of crops if they are intercropped, for example maize and cowpeas together).

This data is largely based on "recall" information – the farmers remember or recall as well as they can how much time they spent preparing a plot (and they have good understandings of the sizes of their plots), how much money (if any) they spent on seeds, fertilisers, or other inputs, how much they harvested from each plot, and how much money they got when they sold it (or, for produce grown and kept on the farm for food later in the season, how much it would have been worth if they had sold it).

From this information, they calculate the profit, or possibly loss, from each crop or crop combination. They also calculate the profit per acre (or hectare), and the profit per man- or woman-day of labour for each crop or combination of crops.

At the end of the three sessions most of the farmers are able to calculate the gross margins for the main crops they have grown, and also the gross margins per acre, the gross margins per man- or woman-day, and the cash outlays per acre for each crop.

This information from all the farmers is then put together. Between them they will have grown a variety of crops, with different kinds of seeds, different purchased inputs, different times of planting, different uses of irrigation where this is possible, and different combinations of crops on a plot, etc. There will also be other differences, such as different soils or varying rights to access water for irrigation. However, most of the farms will be sufficiently alike for it to be meaningful to calculate the average gross margins per acre or per person-day for each crop or activity or combination of crops.

The farmers then discuss which crops to grow in future. The gross profit is not the most interesting figure to them. A few farmers, who have access to irrigation water, can get high returns by growing crops in the dry seasons. Some have larger plots than others. The gross margin per acre made from each crop is more relevant for farms with only a limited amount of land, because this figure identifies how farmers can get the most from whatever land they have. For farmers who are not able to afford to purchase inputs such as fertilisers, the rate of profit (i.e. the gross margin as a percentage of the cash that has been invested to get it) is the most relevant. The gross margin per person-day is relevant for a farm with a limited labour force and no ability to hire extra labour.

With this information, individual farmers consider whether changes would be likely to improve their position, for example whether purchase of a capital asset, such as a power tiller or a well, would be a good investment. The farmers can see that the best solution will not be the same for all farmers – there is no one perfect crop or combination of crops. Combinations are often better than planting pure stands of one crop. They can also consider how they would make the transition – what impact it would have on labour at different times in the year, and what needs there are for cash, as well as land.

From this the farmers can go on to consider their crops together, and to work out a "labour calendar", which will show them how they should use their labour most effectively throughout the seasons, on different crops. In doing this, they are, in a basic way, moving into farming systems (or farm management) studies – without needing to use the complicated language associated with most agricultural research.

Overall, training like this has helped the farmers to better understand:
- The value of profitability analysis;
- The use of indicators (gross profit, gross margins for each crop or combination, gross margins per acre, gross margins per day of labour, and return on money spent).

From that it is easy to draw appropriate lessons:
- That much of what they need to know is already available: on their own farm, or, in the village among neighbours, or through discussion with extension workers or field officers from NGOs or projects, or the owners of shops selling agricultural supplies.
- More is not always better: planting large plots can sometimes bring less cash than planting smaller plots.
- The selection of a crop or crops depends on a farmer's assets and therefore the best solution will be different for different famers. What is best for one farmer may not even be possible for another.

Further reading on open access on the internet

1. Sarah K. Lowder, Jakob Skoet, Terri Raney "The Number, Size, and Distribution of Farms, Smallholder Farms, and Family Farms Worldwide" *World Development 87*, 2017 pp.16–29. http://www. sciencedirect.com/science/article/pii/S0305750X15002703
2. Benjamin E. Graeub, M. Jahi Chappell, Hannah Wittman, Samuel Ledermann, Rachel Bezner Kerr, Barbara Gemmill-Herren "The State of Family Farms in the World" *World Development* Vol.87, 2016 pp.1–16 http://www.sciencedirect.com/science/article/pii/S0305750X15001217
3. Andrew Coulson, "Small Scale and Large Scale Agriculture: Tanzanian Experiences" in Michael Stahl (ed.) *Looking Back, Looking Ahead – Land, Agriculture and Society in East Africa*, Nordic Africa Institute, 2015, pp.44–73. http://www.diva-portal.org/smash/get/diva2:850493/FULLTEXT01.pdf. (An earlier version is D Gabagambi and A Coulson "'Small' or 'Large' Farm? Which Way for Tanzania" in Lucian Msambichaka et al (eds) *How can Tanzania move from Poverty to Prosperity?* University of Dar es Salaam Press, 2015, pp.248–276)
4. Deborah Fahy Bryceson 'Reflections on the Unravelling of the Tanzanian Peasantry' in Michael Stahl (ed.) *Looking Back, Looking Ahead – Land, Agriculture and Society in East Africa*, Nordic Africa Institute, 2015, pp.9–36. On the internet at http://www.diva-portal.org/smash/get/diva2:850493/FULLTEXT01.pdf
5. Michael Collinson (ed.) *A History of Farming Systems Research.* FAO and CABI Publishing, 2000 https://books.google.co.uk/books?id=3OyMszpzu_QC&printsec=frontcover&source=gbs_ViewAPI&redir_esc=y#v=onepage&q&f=false
6. Michael Collinson *Farm Management in Peasant Agriculture.* Westview Press, Colorado, 1983. http://pdf.usaid.gov/pdf_docs/PNAAP952.pdf
7. *Developing Business Skills in Smallholder Farmers as a Basis for Income Growth.* Briefing Paper, Grow Africa Smallholder Working Group, AGRA, 2018 https://www.growafrica.com/groups/briefing-paper-developing-business-skills-smallholder-farmers-basis-income-growth-0

Further reading not on open access on the internet

8. James Scott *Weapons of the Weak – Everyday Forms of Peasant Resistance* Yale University Press, 1985
9. Michaela von Freyhold *Ujamaa Villages in Tanzania: Analysis of a Social Experiment*, Heinemann, 1979
10. Mueller, Bernd "Tanzania's Rural Labour Market: the Unstudied Link between Poverty". In: Carlos Oya and Nicola Pontara, eds. Rural Wage Employment in Developing Countries:.Theory, Evidence and Policy. Routledge, 2015.
11. Frank Ellis *Peasant Economics: Farm Households and Agrarian Development.* Cambridge University Press, 2nd edition 1993.

Topics for essays or exam questions

1. Explain how small farmers can minimise the risks they face.
2. There are at least 15 reasons why small farmers often plant more than one crop in the same field. Explain the logic behind as many of these as you can (up to about 10!)
3. Why are systems of mixed cropping (planting more than one crop in the same field) more commonly found on small farms than on large? (Hint: many of the reasons relate to the use of machinery)
4. Give, with explanations, three different reasons why family labour is often more efficient than wage labour.
5. "Farming is a business. But it is often a business where there are many objectives, not just making the most money in an average year." Explain this statement.
6. How would you expect farmers to react if they get very low prices for their crops? Give some illustrations to support your arguments.
7. Discuss the uses of the words "commercial" and "non-commercial" in connection with small farmers. What are their main objectives? Explain why it makes sense to describe them as commercial.
8. "It is extremely unhelpful to call small farmers 'peasants' if that is intended to imply that they are inefficient." Discuss the difficulties involved in using the word "peasant"?
9. If small farmers are ordered to take some action, but they do not believe that this is in their best interests, how are they likely to react, and why?
10. "Calculations which consider each crop separately, such as Gross Margin Analysis, are often only a starting point for small farmers who can use only the land and labour that they have to grow a variety of crops". Discuss this. What conclusions can you draw?

CHAPTER 6

Large Farms

Key themes or concepts discussed in this chapter:

- The different kinds of large farms and their ownership.
- The capital costs of investing in large farms.
- The economics of large farms – recurrent costs, income streams, profitability, risks.
- The productivity of large farms compared with small.
- Economies and diseconomies of scale in agriculture.
- The consequences of very close links with processing and export markets.
- Agribusiness – and how it makes profits from the sales of inputs and seeds.
- Situations where large-scale farming in Africa should be encouraged.
- Land tenure, and the availability and acquisition of land for large farms.

WHAT IS A LARGE FARM?

As shown in Chapter 5, there is no unambiguous definition of a small farm, but it is one with sufficient available land for a family to grow most of its food and to produce crops or other agricultural products that are sold to get the cash needed for day-to-day existence.

A large farm is the opposite: it uses more labour than is available from a family, and has more land than is needed to grow food for the family and those who work on it.

Most of what was said in Chapter 1 about soils and their chemistry, and about mechanisation and conservation agriculture in Chapter 2, applies to farms of any size, as does the discussion of livestock. Many of the cautionary tales about irrigation in Chapter 3 relate to farming on a large scale. Much of the discussion of contract farming in Chapter 8 is highly relevant for large farms. However, this is the only chapter that considers large farms explicitly.

The chapter starts by categorising different types of large farm, and their ownerships. It then looks at the economics of large-scale farms, and the risks they face. That leads to a discussion about how small farms may succeed in competition with large, and, in many situations, produce higher average yields. This leads in turn to the issues raised by different forms of land tenure, and especially to how large farms may acquire land. Is this, as some would argue, "land grabbing"? Or is some kind of increased scale of farming inevitable if higher productivities are to be achieved? The chapter concludes that there is a case for large-scale farming, but not where it conflicts with small-scale.

LARGE FARMS: A CATEGORISATION

The ownership, and the management, of large farms can differ. Here is a simple classification:

Most **plantations**, or **estates**, include processing factories, such as for sisal, tea, sugar or coffee. These are expensive to construct and require specialist management and engineering support. Most are set up by companies with backing from private shareholders and/or financial institutions. Many of their agricultural operations are labour intensive, especially the harvesting of the crops, so most plantations include labour forces which have to be trained, managed, provided with housing, and paid and employed with formal contracts. The owners may decide to supplement production from their own "nucleus farm" by contracting outgrowers. This can increase the supply of the crop, make full use of the processing facilities, and avoid the inflexibilities and overhead costs of employing workers directly.

Most, but not all, plantation companies in Africa are externally owned. An example of a locally owned company is the Karimjee Jivanji group, which started as a hardware shop in Zanzibar in 1818, became a trading company, and much later the biggest and most progressive owner of sisal plantations.[7]

Settlers are families from another country, mainly from Europe, who came seeking a new life. Most arrived with limited resources, and grew crops that did not involve expensive processing – such as coffee, maize and wheat, or vegetables such as beans or potatoes; some kept dairy cattle. As far as they could, they built their own houses and maintained their own machinery. Banks were often reluctant to lend to them, unless they offered the farm as security; but that would mean that the farmer would lose the land if anything went wrong. In colonial times, a few settlers had good land, chose the right crops at the right time, and became rich. For many others life was a struggle and they looked for support from the governments of the time. Some became involved in politics, and, where there were many settlers, as in Zimbabwe, Mozambique or Kenya, their political influence was significant. Other colonies, such as Uganda, did not allow settlers.

Large-scale African farmers face many of the same problems as settlers – unless they have access to external money or savings, or can get preferential treatment from banks. At the start of this book we noted that many who made savings during time spent in politics or the civil service have been able to get access to land and to invest in agriculture.

A **cattle ranch** depends on the grazing being controlled, so that the grass flourishes and the cattle are well fed throughout the year. Cattle need reliable supplies of water. They may also need supplementary feeding to get through the dry season. The necessary fencing is expensive to construct and difficult to maintain – so a ranch is not a cheap way of raising cattle.

Intensive **poultry, pig or fish farms**. These farms do not require large areas of land. Much or most of the feed is grown elsewhere, and either prepared and mixed in factories, or, if supplies are available locally, on

---

[7] In 1955 the family donated "Karimjee Hall" to Dar es Salaam City Council; it was used for meetings of the Tanzanian Parliament until a new building was constructed in Dodoma. When Tanzania became independent in 1961, the Speaker of the Parliament was a member of this family.

the farm. There are health and animal welfare issues with these farms. To keep chickens or grow fish in artificial ponds free from disease, regular doses of antibiotics may have to be used. These may get into the food chain when the meat or fish is sold. If the animals are unhealthy because they do not have sufficient room to move, and get no exercise, or because their living conditions are not kept clean, more antibiotics are likely to be used.

State farms. Governments often own or acquire large farms by nationalisation, or create them on unused or under-utilised land. Sometimes they have inherited the assets of investments which failed, such as the Groundnut Scheme in Tanzania after the Second World War, or they may have acquired land from settlers who left their farms after wars or political change. Forest plantations are frequently state-owned. The ownership of land that can be irrigated from large state (or donor) investments is often vested in the state, even when the actual farming is undertaken by individuals.

Agribusiness. The word "agribusiness" is often used to refer to a farm developed by a large multinational corporation, often with other interests – for example the production or distribution of fertilisers, seeds, or food. Such a farm may have a contract to supply specific quantities of a crop such as peas for freezing or canning, to exacting quality standards and at specified points in time. The marketing is built into the business plan from the start; but the possible downsides of contract farming discussed in the previous chapter can still be relevant.

Agribusiness in Africa is not new. The tea merchant Brooke Bond developed many of the Tanzanian tea estates. The Commonwealth Development Corporation established a large-scale wattle plantation near Njombe in Tanzania in 1949, as well as tea plantations in Kenya, Uganda and other parts of Tanzania, using the most advanced agricultural techniques of the time. So did the farms of the Basuto Wheat Scheme, which set out to grow wheat in northern Tanzania in the late 1970s, but never made a profit, despite receiving the best available advice, and assistance from the Government of Canada (see Chachage [1] for more detail).

Another use of the term "agribusiness" is to refer to highly capital intensive, often indoor, agricultural production" – for example the cultivation of flowers such as roses and lilies, vegetables such as tomatoes, and huge sheds where dairy cows never leave their stalls. This type of "factory farming" has been developed in the Netherlands where crops are grown in what can accurately be described as factories, with maximum use of science and technology, and minimum employment of labour. From there the technology spread to South America, supplying US markets, and to Japan. Some products grown under contract elsewhere, for example roses in Kenya or northern Tanzania, feed into this kind of system, which can supply very large quantities of uniform products, of guaranteed quality, at low prices, for long periods of the year.

Where these systems exist, it is almost impossible for other producers to compete. They are, however, extremely expensive to set up and to maintain, and they need constant attention to deal with plant and animal diseases. The intensive production of livestock takes little account of the welfare of the animals, and risks the health of those who work in these conditions, especially from exposure to pesticides. There are also big downsides from the use of antibiotics to control infections in animals and birds such as chickens, in conditions under which resistance to those antibiotics is extremely likely to develop. These matters are discussed further in Chapter 10.

COMPARISON OF LARGE AND SMALL FARMS

Large farms may be highly productive in terms of output per unit of labour and yields per unit of land area. However, and this is surprising to many people, many empirical studies have found that yields per hectare from small farms are, on average, often higher than those of large ones in the same area.

This "inverse relationship" sometimes occurs because large farms neglect or do not fully develop much of the land allocated to them (Ellis, **Chapter 4 [15]**). This was true with many settler-owned farms in Zimbabwe prior to the land reforms of the 1980s and beyond, which divided the land among African farmers. This suggests that a government should not allocate more land to a farmer than he or she can manage, and should retain the right, if that land is not developed within a few years, to take it back.

The inverse relationship also occurs because small farmers use the techniques described in Chapter 5, in particular multiple cropping, to farm the land intensively – often this is necessary for them to survive. The amount of (unpaid, family) labour per unit of land is much higher than could be afforded on a large farm, and the productivity per hectare is as high or higher when all the different crops, and the incentive which comes with farmers working on their own land, are taken into account.

Studies which document the inverse relationship are usually country-based, and the data is not always comparable – for example, in some of the Indian studies, "large" is anything bigger than 5 hectares; in other studies "large" is much more than that. The inverse relationship is not an absolute statement or law, true in all situations. There are certain crops and activities – such as the mechanised production of wheat described in the case study at the end of this chapter – where the economic advantages of large scale are difficult to dispute. The inverse relationship can be considered as a warning: do not dismiss small-scale farmers without considering how they may compete with large farms.

THE ADVANTAGES OF LARGE SCALE FARMING

So what are the advantages of large farms?

1. **Finance.** When it comes to agribusiness, there should not be a contest. If an economic case can be made to the shareholders and their bankers, large farms should be able to access the resources needed to develop their land, to gain access to machinery and to chemicals, to follow the latest research, and to use the latest technologies. Where capital is readily available, large farms should be able to get higher yields than small ones, even though, as just noted, in reality this is often not achieved.

2. **Marketing.** Large farms have special advantages when it comes to marketing, as shown in the next two chapters. A large farm owned by a company that also owns processing facilities, or that can work closely with a processing factory, and can produce and sell in large quantities, has advantages over a small farm. It can more easily satisfy the certification and quality standards required to gain access to overseas markets. It can meet the terms of contracts with processing companies or exporters. It can expand its production relatively quickly if there is a demand from those who process the crop.

3. **Procurement.** Large firms can obtain inputs and machines more cheaply than small-scale users. They can ensure that they get good quality seeds, fertilisers, etc., store them safely, and use them appropriately. They are better able to maintain agricultural machinery and equipment.

4. **Economies of scale.** Most agricultural technologies are scale-neutral. Thus fertilisers work just as well on a small farm as on a large one. But this is less true of certain types of agricultural machine. Thus a combine harvester requires a large area to justify

its cost. So does a very large tractor, and the equipment designed to draw on its power. These machines were developed in countries where labour is very expensive. Where labour is cheap, the gains are much less, and in remote parts of Africa the costs of transporting large machines, and maintaining them, are much greater than in, say, the plains of the US Midwest.

5. **Processing.** Crops which deteriorate if left for more than a few hours (such as tea or milk), or because unprocessed they are bulky and expensive to transport long distances (such as sugar or sisal), require processing in factories close to where they are harvested and are especially appropriate for large farms. It may nevertheless suit the owners of these factories to outsource some of the production to outgrowers. The case study of Tanga Fresh at the end of Chapter 8 shows that small farms can economically supply fresh milk for urban markets. However, cut flowers (such as the roses and chrysanthemums now produced in bulk near Nairobi and Arusha) have to be picked at exactly the correct time, handled with great care, packed under factory conditions, and air-freighted to wholesale markets in the Netherlands or elsewhere. They are difficult to grow using outgrowers, and are better suited to large farm production.

THE DOWNSIDES OF LARGE-SCALE FARMING

1. **Establishment costs.** Establishing an arable farm on virgin land is seldom quick or cheap. The bush and most of the trees have to be cleared. If tree roots and rocks remain, they can destroy machinery. If the land is uneven it may need to be levelled. The land will need to be fenced, especially on livestock farms. Soils have to be protected against erosion. Watercourses need to be conserved. If irrigation is required, pumps, canals and drainage channels have to be constructed. Access roads and all-weather bridges are needed. Also houses, workshops, stores for spare parts and storage areas for harvested crops. If processing is to take place on the farm, the specialist machines will need to be installed, and staff trained to operate and maintain them. Even if a large farm has already been developed when it is acquired by a new owner, rehabilitation of all these assets is usually required and is a very costly process.

2. **Working capital requirement.** Once the initial establishment costs have been financed, a new farm will need working capital – to cover costs while the crops are growing and before they are harvested.

For tree crops this can mean waiting several years until there is any income. Once the farm is running there are ongoing overhead costs, including the cost of salaried managers, of training and employing a work force, maintaining machinery, and purchasing seeds and chemical inputs.

3. **Vulnerability to risk.** Large farms are vulnerable to most of the same risks as smaller farms: insufficient rainfall or water, or too much water when it is not needed, an unexpected plant disease, an attack by birds or locusts, the breakdown of a machine, the loss of key personnel, or an unexpected fall in the price of the product. Thus in Tanzania in 2012, large farmers as well as small were affected by a sudden drop in the price of rice when the government unexpectedly authorised large imports to meet the demands of urban consumers. Similar problems were experienced when sugar imports were permitted. With heavier investment in their land, capital assets and labour force, large farms are at greater risk compared to small farms of suffering financial collapse when things go wrong.

4. **Inflexibility.** Large farms have to continue to pay their overhead costs, including the wages and salaries of labourers and managers – in bad times as well as good. This inflexibility is one reason why much of the technology on large farms in Western countries has been developed to replace labour. Where machines cannot easily do the tasks, such as harvesting many fruits and vegetables, labourers are often hired just for the few weeks that they are needed, but this is rarely an ideal solution.

5. **Management and Employee loyalty.** Management of a large farm is a complex affair, and in most countries skilled managers are in short supply. Furthermore, compared to small farmers working on their own land, hired labourers, especially if employed on a short-term basis, are likely to have no long-term loyalty to the farm, and to cut corners or do less work if they can. So there are high costs of supervision. To counter these tendencies, some large farms seek ways of giving their employees a stake in the profitability of the enterprise, so that they have the same incentive as small farmers to make "their farm" succeed, but this is not easy to achieve.

6. **Subsidies.** Large-scale farmers frequently lobby governments to subsidise them, either through guaranteed prices or by asking the government to pay for construction of infrastructure, such as roads, dams or social facilities. Farms in the United States were subsidised to keep them open during the depression of the 1930s,

and they are still subsidised today. The European Union was built around a "Common Agricultural Policy" which involves extensive subsidies paid to farmers. These subsidies enable governments to keep part of their population engaged in work in rural areas, and to ensure reliable supplies of food at relatively cheap prices, but apart from the cost of the subsidies they result in unfair competition with producers in other countries. There is now strong pressure from governments in South America and Africa to persuade developed countries to remove or reduce these subsidies. This has had some success: under pressure from the World Trade Organization, the USA has reduced its subsidies to cotton farmers, though with increased use of synthetic fibres this has not led to a significant increase in world prices for cotton.

Given all these issues, it is not hard to see why large-scale farming is not always profitable, and why the owners of large farms often use outgrowers rather than expand production themselves. An example of a farmers' co-operative established in Tanzania soon after independence, which attempted with some success to realise the benefits of large-scale farming and to overcome the downsides, is described in the case study at the end of this chapter.

LAND TENURE

Land for agriculture, the rights of the users and the powers of the state to acquire land and reallocate it for other uses, is controversial in every country and leads to frequent disputes which are often difficult to resolve.

Whether farms are large or small, the users require secure land tenure to give them the confidence to make long-term improvements and to use the land sustainably. The form of tenure may be *freehold* (where the farmers have title to the land in their own names) or *leasehold* (where another person or an organisation such as the national government or a village government holds the title but guarantees to allow the farmers to keep using it).

Before colonial interventions, land in many parts of Africa was held by ethnic groups or clans. Colonial governments, as part of their "indirect rule", created chiefs in areas where these did not already exist. These governments then gave various powers to chiefs, including powers over allocation of land. So if a family needed land, the chief would be expected to find some, and, conversely, if for any reason a family was not using its land, it could be taken away. This "customary law" was recognised in courts. Where possible, it incorporated practices from pre-colonial times. There were different arrangements for land in urban areas. Settler farms and plantations and land used by the state, such as for game reserves or prisons, were surveyed and mapped and given "titles" as "alienated land".

As an illustration of the difficulties, consider the situation in Tanzania. After Independence in 1961, Tanzania transferred the ownership of all agricultural land, with few exceptions, to the state. Broadly, rights of occupancy were given to the users, but the legal titles given to settlers in the colonial period continued to be recognised. In 1999 the government passed two new laws.

1. The *Land Act of 1999* defined two types of land. "Reserved Land" consisted of national parks, forest and game reserves, and other land used by the government for specific purposes. "General Land" included surveyed land in urban areas, and "unoccupied or unused" land in rural areas which the government could easily acquire and reallocate for other uses.

2. The *Village Land Act of 1999* defined a third category, "Village Land". This covers 70% of Tanzania's land area and includes almost all the land used by small farmers. The Village Land Act gave land use rights to individual farmers, and gave Village Councils, acting on behalf of Village Assemblies, responsibility for surveying Village Land, administering it, and registering its ownership. If Village Land is allocated for some other use, such as to an incoming investor, fair compensation must be paid.

The 1999 Acts have proved controversial, firstly because there are contradictions between the provisions of the old law and the new, but also because of the impracticality of implementing many of the provisions. There is no realistic prospect of large areas of village land being registered, since most villages lack the expertise to carry out the necessary surveys and complete the required documentation. There is no clear definition of "unoccupied or unused" land – which is clearly ambiguous when land is used for shifting cultivation or nomadic pastoralism. This has left much land open to being transferred to new owners, including land left unused when farmers were forced to move into villages in the 1970s. When women separate from their husbands or are divorced, their rights to own land are not always recognised (see Da Corta and Magongo [2]). The powers given to the central state are also controversial, especially when a company or a large farmer is seeking a large area of land. The 1999 reforms were intended to give power to village councils over land in their villages, but they also made it possible for central government to overrule the councils or to take land that was not clearly registered as village land, designate it as general land, and allocate it to someone from outside for a different purpose (see also Sundet [3]). In 2016 a new Draft Land Policy was proposed to clarify these issues, but to date it has not been put into effect.

An argument often used for giving individual farmers titles for land, whether as freehold or under a long lease, is that banks can then use the land as security – if a loan is not repaid then the bank can take possession of the farm. The problems that this can create for small farmers are considered in Chapter 8, where lending to co-operatives or farmer groups is suggested as an alternative. Clearly large farms need titles. The real question is how much the state needs large farms, and how much priority it chooses to give to large farmers and foreign investors over small farmers.

LAND ACQUISITION – OR LAND GRABBING?

The high prices of both oil and food grains on world markets from around 2008 onwards, and the apparent availability of land suitable for agriculture, made African countries a target for investors from Europe and America, and from Asia and the Middle East. In 2009 Sulle and Nelson produced a table [4, pp.16–17] showing 280,000 hectares "allocated" to seventeen such projects in Tanzania. In 2013 Locher and Sulle updated the data, having checked as many as possible of the projects on the ground [5]. Their work showed that promises had been made or land transferred for more than 375,000 hectares (plus another 325,000 hectares in the far west of Tanzania which was being farmed by refugees from Burundi and elsewhere), though many of the proposed projects had not proceeded beyond press releases which indicated that an investor was in negotiation to gain access to a large area of land. Several investors who had set out to grow biofuels had made slow progress or failed.

In 2012 the Tanzanian policy statement *Big Results Now* identified 25 "commercial farming deals" for paddy and sugar. Two of the farms were to be larger than 50,000 hectares, 9 were to be "nucleus farms" of over 20,000 hectares, 5 of the remainder over 10,000 hectares, and all but two of the rest 5,000 hectares or more. To get a feel for these sizes: 20,000 hectares is the size of a square where each side is roughly

14 kilometres (close to 9 miles) in length; 50,000 hectares is a square over 22 kilometres each way. Even in the plains of North America and Australia there are few farms of this size.

There is little current information on the progress of these proposals. The costs of investment, the logistical challenges, the difficulties of employing and managing large labour forces, and the uncertainties of markets and marketing, have all proved difficult. *Big Results Now* was officially suspended by the President in 2017.

CONCLUSIONS

In an influential review of a number of African countries, the World Bank economists Deininger and Byerlee [6, p. xlii] concluded that:

> The ... evidence suggests that large-scale expansion of cultivated area poses significant risks, especially if not well managed. As the countries in question often have sizable agricultural sectors with many rural poor, better access to technology and markets, as well as improved institutions to improve productivity on existing land and help judiciously expand cultivated area, could have big poverty impacts. Case studies illustrate that in many instances outside investors have been unable to realise this potential, instead contributing to loss of livelihoods. Problems have included displacement of local people from their land without proper compensation, land being given away well below its potential value, approval of projects that were only feasible because of additional subsidies, generation of negative environmental or social externalities, or encroachment on areas not transferred to the investor to make a poorly performing project economically viable.

> Many countries with large amounts of currently uncultivated land suitable for cultivation also have large gaps between potential and actual yields. Thus even without any expansion of cultivated area, large increases in output and welfare for the poorest groups could be possible through efforts to enable existing farmers to use currently cultivated land more productively.

Tanzania needs a balance between small farms and large. The small farms can be a cost-effective means of producing agricultural products where labour intensive methods are appropriate, and where marketing is straightforward. The large farms may be able to break into some overseas markets that would otherwise be difficult to enter (such as cut flowers – though seeds for the flowers can be produced by small farmers). But these large farms would employ few or relatively few workers, most of them on an insecure basis, so they would make only a small impression on poverty reduction. What happens in its *small* farms will therefore be crucial for Tanzania as it develops, and in particular for its ability to feed itself. Care needs to be taken to ensure that allocation of land for large-scale investments does not reduce the amount of land available for small farmers, and put realisation of their potential at risk.

Case Study 6: Upper Kitete Co-operative Farm

This case study describes the experience of a large scale co-operative farm established in Northern Tanzania in the early 1960s, which is still operating today, albeit in a radically different form. It illustrates many of the issues raised in Chapter 2 regarding the benefits and risks of mechanised agriculture, as well as the issues discussed in Chapter 6 regarding the economies of scale that can be realised on large farms. It is written by Antony Ellman, who was the first Manager of Upper Kitete Co-operative Farm from 1963 to 1966. He has revisited the farm and watched its progress at frequent intervals since then.

Tanzania's transformation strategy

When Tanzania gained independence from Britain in December 1961, the new government was determined to transform smallholder agriculture, primarily through the widespread use of tractors and high-yielding varieties of crops and livestock. This "transformation strategy" was closely linked to a plan to persuade farmers on scattered smallholdings to move into villages, where the government could more easily supply services such as piped water, schools and medical facilities.

At the same time the farmers would be encouraged to form producer co-operatives, through which they would realise economies of scale in using machinery, obtaining finance and marketing their produce. The producer co-operatives were seen as the first step towards transforming rural communities into the socialist society which the government hoped Tanzania would one day become.[8]

With advice from the World Bank and a loan from the British Government, a programme of 23 model "village settlements" was initiated in different parts of the country to spearhead this transformation. The majority of the new settlements were poorly planned and managed, and closed down within a few years. However, one scheme was highly successful in its early years, and 55 years later it is still in existence.

---

[8] The Tanzania Government's attempts to create socialist communities in the rural areas were embodied in the "Ujamaa Village" programme of the 1960s and 1970s, in which farmers were told to move into villages and to farm at least part of their land collectively. It was hoped that this programme, in addition to facilitating provision of community services, would allow the potential benefits of large scale farming to be realised without creating extreme inequalities between large and small scale producers. The programme was abandoned in the 1980s, when it was seen that the social and economic objectives were not being realised. The Upper Kitete case illustrates some of the lessons learned. See also M von Freyhold (Chapter 5, Further Reading, Reference 9).

Upper Kitete Co-operative Farm

This farm occupies a 6,500 acre (2,600 hectare) block of highly fertile land in the northernmost tip of Karatu District, some 100 miles west of Arusha. The area, situated between Ngorongoro Forest Reserve and the Rift Wall, was claimed by two ethnic groups, the pastoral Maasai and the agricultural Iraqw, but it had been kept largely unoccupied during the colonial period as a "no man's land" to avoid the risk of conflict between the two groups.

In 1963 Upper Kitete was populated primarily by wild animals – buffalo, elephant, leopard and lion – though a few intrepid Iraqw farmers had started to grow wheat in the area, and many others were eager to do the same. There was minimal infrastructure (very bad roads, no water, housing, health or education facilities). The government decided that it was an ideal location for the transformation of agriculture through carefully planned and managed resettlement.

Upper Kitete has fertile volcanic soils and good rainfall, which allows two harvests of fast-growing crops to be obtained per year. Mechanised cultivation of wheat and barley, linked to intensive production of food crops, vegetable growing and livestock rearing, were the enterprises that were deemed most suitable for the new village settlement. To realise economies of scale, particularly with cereal crops which are difficult to grow without mechanisation, and to guard against the threats of animal attacks as well as Maasai raids, it was necessary, at least initially, for the major part of the farm to be managed as a single large-scale unit.

Rather than transfer the land to a company or to a single farmer, which would have been politically completely unacceptable, the government decided to allocate the land to a hundred landless or very small-scale farming families from the district, who would hold and farm it as a co-operative. In line with the country's socialist goals at the time, it was decided that Upper Kitete should be run as a single collective farm along the lines of an Israeli kibbutz9, despite the novelty of this approach and uncertainty about whether it would be locally acceptable and whether it could result in a viable economic enterprise.

In fact, there were large numbers of landless and small farmers to the south of Upper Kitete, who had long had their eyes on the fertile land

---

9 A kibbutz is a democratically managed collective farm in which members provide their labour and receive an equal share of the profits. It was pioneered in the early years of the Israeli State. A related model is the "Moshav Shitufi", in which the bulk of the land is collectively held and farmed, but each member family also has an individual plot for a house and a homefood garden. The Moshav Shitufi is closest to the model initially adopted at Upper Kitete.

that was available in this land-scarce part of the country. They proved ready to accept the collective model of development that was proposed. Indeed they would have accepted almost any model which would give them access to this prime agricultural land.

The early years, 1963–1966

The opportunity to farm at Upper Kitete was widely advertised around the District, and a hundred farmers were selected out of some five hundred who applied to become members of the co-operative farm. Each of the selected families was allocated a 3-acre homestead plot in the central village, on which to build a house and grow food crops. A plan was drawn up for running the rest of the land as a single unit, with some 2,500 acres allocated for mechanised cereal cultivation, smaller areas designated for irrigated vegetable production, and the rest of the land reserved for grazing.

Machinery was purchased with the loan from the British Government – five tractors with ploughs and harrows, two seed drills, a crop sprayer and a combine harvester. The farmers began work clearing the land, as well as building their houses and planting food crops on their homestead plots. Five people were hired for specialist jobs that the co-operative members could not do on their own – a workshop manager, two drivers, an accounts clerk and a storekeeper. In addition a Settlement Manager and an Agricultural Extension Officer/Assistant Manager were seconded to the co-operative by the government. All other jobs were undertaken by the farmers themselves and their families.

To plan and organise their work, the farmers elected a Village Development Committee, and formed work groups of ten, each with an elected group leader. Some co-operative members who had skills as tractor drivers, builders and livestock herders were assigned to these tasks and given supplementary training. The others were allocated tasks clearing the arable land, constructing access roads and bridges, digging the trench for a piped water supply, building a workshop, store, staff quarters, and later on a primary school and village dispensary.

The soil at Upper Kitete is highly erodible, so broad-based terraces were constructed across the slopes, using heavy machinery (bulldozer and grader). This was a costly process which runs counter to the conservation farming technology described in Chapter 2; its importance had not been appreciated at the time. However, the terraces were effective, and there has been very little soil erosion at Upper Kitete.

Between 1963 and 1966 the progress made by the co-operative members was extraordinarily rapid: 2,000 acres of land were cleared, a hundred houses, stores, workshops, a school and dispensary were built; 1600 acres of wheat and 300 acres of food crops were planted; cattle and sheep were purchased and farmers were trained in technical and managerial skills. High yields of wheat and barley were harvested and sold, family nutrition and household incomes improved. Despite the inevitable difficulties of managing any co-operative enterprise, the community was broadly united, had paid back part of its initial loan, and was moving towards achieving economic viability and successful self-management.

The middle years, 1966–1970

Once the establishment phase was over, with the land cleared and brought into cultivation, the demand for the farmers' labour was greatly reduced. Essential buildings had been constructed, the infrastructure had been improved, and there was a risk that the farmers would become passive bystanders while most of their land was cultivated by machinery.

This risk was countered to some extent through diversification into more labour-intensive activities: irrigated vegetable cultivation (though the available water was insufficient to do this on a large scale), agro-processing (maize and oilseed mills were established), carpentry and clothes-making businesses (this last making fuller use of the women's skills).

However, the contradiction between mechanised agriculture on the one hand, and growing population pressure on the land on the other, was becoming increasingly acute. With the benefit of hindsight, the adoption of more labour-intensive production techniques, using for example ox-ploughs rather than tractors (as had been practised by many of the farmers before they moved to Upper Kitete, and indeed by some on their 3-acre homestead plots), might have been a more appropriate choice.

The villagisation campaign, 1970–1980

These contradictions became even more acute after 1970, when the government adopted a programme of villagisation throughout Tanzania. The Upper Kitete Co-operative had 6,500 acres, shared between one hundred families (thus notionally 65 acres per household), while the average farm size per household in other parts of the District was less than ten acres. To redress this imbalance, the government allowed 350 new families to enter the scheme.

There was minimal consultation with the original members of the Upper Kitete Co-operative, and this imposition inevitably caused tremendous resentment. Just as the 100 pioneer families were beginning to enjoy the fruits of their labour, they were asked to share the benefits with 350 new members who had contributed nothing to the development phase. Half of the collectively held land had to be distributed to the newcomers as homestead plots, and the land that remained was too small an area to justify maintenance of the machinery that had been acquired and the management system that had been developed.

With smaller holdings to manage, many of the farmers and the Co-operative itself switched from growing wheat and barley, which are best grown with tractors and combine harvesters, to more intensively produced crops like maize, beans and other vegetables. The collective management system needed for running the farm as a single unit was no longer appropriate, so the social cohesion of the community inevitably became weakened.

The current phase, 1980–2015

Over the last thirty-five years much of the land that was collectively owned and farmed by the Upper Kitete Co-operative has been divided among the individual members. The machinery that belonged to the Co-operative has fallen into disrepair. Some of the land that remains with the Co-operative has been leased to large-scale farmers from elsewhere in the District, who continue to grow cereal crops using full mechanisation. But for the most part the production system has changed from large-scale mechanised agriculture to the more labour-intensive production of higher-value crops and livestock.

The approach to soil conservation has also changed, with many of the elements of conservation agriculture described in Chapter 2 being adopted on the smaller land holdings. The broad-based terraces, constructed at high cost in the early stages of the scheme, did protect the soil effectively against erosion, but they could not easily be maintained once the land had been divided up. The alternative approach to soil conservation described in Chapter 2 – minimum tillage, keeping the soil covered, and rotating between a variety of crops – is proving more sustainable on the small-scale holdings.

The population at Upper Kitete has grown from around 500 in 1963 to over 5,000 today – a tenfold increase in the fifty-five years of the Co-operative's existence, roughly twice the rate of population growth in the country as a whole, due largely to the introduction of new members.

Even with more intensive use of the agricultural land and diversification into non-agricultural economic activities, it is not easy for such a large population to achieve an acceptable standard of living off a limited land area. Many of the younger people have now been obliged to move to other parts of Tanzania or beyond, in search of land or employment.

Analysis of the experiment

The Upper Kitete Co-operative represents a bold experiment which, in its early years, succeeded in combining the strengths of smallholder farming – particularly the incentive that comes when farmers work on their own land – and the economies of scale that can be realised on a large-scale farm. The rapid progress made in the first three years could not have been achieved had the land been divided into smallholdings, nor even if it had been given to a single large farmer – in view of the lower motivation of hired labourers, and of the antagonism that would have been created among the people excluded from the land.

As the population dependent on the land expanded, partly through natural increase but particularly through the introduction of additional families during the villagisation campaign, the production technology and the management system that were appropriate had to change to more labour-intensive and individualised approaches. The economies of scale achievable through large-scale farming were lost, and the social cohesion of the community was weakened.

The challenge for the future will be to find production systems which are sufficiently intensive to support the greater population, and to recreate the solidarity and enthusiasm that enabled the community to make such remarkable progress in the early years.

Further reading on open access on the internet

1. Chambi Chachage *The State of the then NAFCO, NARCO and Absentee Landlords' Farms/Ranches in Tanzania*. Land Rights Research and Resources Institute (LARRI/Hakiardhi) 2009 https://www.researchgate. net/publication/308796270_The_State_of_the_then_NAFCO_NARCO_ and_Absentee_Landlord's_FarmsRanches_in_Tanzania

2. Lucia Da Corta and Juanita Magongo *Evolution of gender and poverty dynamics in Tanzania*. London: Chronic Poverty Research Centre, Working Paper 203, 2011. https://www.gov.uk/dfid-research-outputs/ evolution-of-gender-and-poverty-dynamics-in-tanzania-cprc-working-paper-203

3. Geir Sundet *The 1999 Land Act and Village Land Act*: A technical analysis of the practical implications of the Acts 2005. http://mokoro.co.uk/wp-content/uploads/1999_land_act_and_village_land_act_tanzania.pdf

4. Emmanuel Sulle and Fred Nelson *Biofuels, Land Access and Rural Livelihoods in Tanzania*, International Institute for Environment and Development, 2009. http://pubs.iied.org/pdfs/12560IIED.pdf

5. Martina Locher and Emmanuel Sulle *Foreign land deals in Tanzania: An update and a critical view on the challenges of data (re)production*. The Land Deal Politics Initiative Working Paper 31, 2013. http://www.plaas. org.za/sites/default/files/publications-pdf/LDPI31Locher%26Sulle.pdf

6. Klaus Deininger and Derek Byerlee *Rising Global Interest in Farmland: Can It Yield Sustainable and Equitable Benefits?* World Bank, 2011. https://siteresources.worldbank.org/DEC/Resources/Rising-Global-Interest-in-Farmland.pdf

Topics for essays or exam questions

1. The inverse relationship – yields from small farms being higher than yields from large farms in the same area – is not what one would expect. Discuss at least two ways in which this can occur.

2. "Most agricultural technologies are scale-neutral." Discuss this statement with relation to the technologies usually associated with the Green Revolutions that took place in Asia in the 1960s and 1970s.

3. "The willingness of sugar, tea and sisal estates to purchase from outgrowers suggests that they can produce these crops more cheaply than the estates." If this is the case, why should these companies continue their nucleus estates? Why not just use outgrowers?

4. Discuss the feasibility of groups of farmers forming a co-operative or producer group to realise the economic benefits of large-scale farming.

5. Discuss the value of subsidies for large-scale agriculture. Under what circumstances can these be justified?

6. Why do many policy statements by governments state that there is plenty of land available for large-scale farming, when in reality it is very hard to find suitable large sites?

CHAPTER 7

Market Access and Value Chains

Key themes or concepts discussed in this chapter

- The role of the marketing system is to increase the quantity of agricultural products marketed and to provide outlets for this produce.
- The marketing of agricultural products begins at the farm, and ends with a consumer. It may be a single transaction, or a complicated chain that includes storage, transport, grading, processing, packaging, and sales.
- An appropriate research technique is to study the value chain. A "value chain analysis" tracks the prices paid for a crop as it moves from farmer, to trader, wholesaler, processor, and eventually to a consumer. It shows how value is added at each stage of the chain. If the intermediate stages can be made more efficient, then potentially there is more value for the farmers, i.e. higher prices.
- Agricultural marketing can be undertaken by a variety of institutions in the private sector: traders, farmer groups, processing or manufacturing companies.
- Publicly owned co-operatives and marketing boards may also be involved, as well as banks, if they provide credit for agricultural activities.
- Many aspects of marketing and processing need to be regulated and regularly inspected, to ensure that high quality is maintained, and that farmers are not cheated.
- Accreditation schemes, which guarantee quality and give access to niche markets are an important means by which farmers can get higher prices.
- Mobile phones offer new ways of providing information, especially about prices.

ROUTES TO MARKET

Chapter 5 stressed how small farmers need to minimise, or at least contain, the risks they face. It also showed how small farmers respond to the prices they receive when they sell their produce, generally planting more in seasons following a price rise, and less when prices fall. Thus marketing is all-important. If the marketing system for a crop does not work well, or if farmers are paid less than they expect, they are unlikely to increase their cultivation of that crop.

This chapter sets out the different ways in which a crop can be sold. That leads to the concept of a "value chain", which identifies the value of a crop at different stages of its marketing. The institutions involved in agricultural marketing, in both the private and the public sectors, are described. The next section is a discussion of accreditation schemes through which farmers can get higher prices for crops where quality is guaranteed or, for example, where crops have been grown without the use of inorganic chemicals or following fair trade principles. The final section examines some of the means through which farmers can get up-to-date information about prices and market opportunities, especially with the assistance of mobile phones.

Chapter 8, which follows this one, is about finding finance for small farms. Much of this will come in the form of credit, from banks or other financial institutions. This is intrinsically linked to marketing, since if crops are not well marketed there will be less money to repay any loans that producers have taken, and they will not be eligible for further credit.

THE MARKETING CONCEPT

The marketing of agricultural products can be a simple matter, as when a farmer sells a small surplus at the roadside or at a small village market. There is no specialisation and the marketing functions are all carried out between the producer and the consumer. A developed marketing system on the other hand is a complex affair, with large investments in roads, equipment, storage depots, shops and personnel. If consumer demand is to be satisfied and if economic growth is to proceed unchecked by high food prices and the importation of food, the agricultural marketing system must not only move goods from the producer to the consumer at minimum cost, but also stimulate production of food for the market, and provide the organisation which will respond rapidly to consumer demand.

To illustrate this, here are some scenarios, of gradually increasing complexity, for the sale of tomatoes:

1. A farmer places a bucket of tomatoes at the side of a main road, and hopes that someone will stop to buy them. He or she may sit behind a table and wait for customers. There is no other infrastructure or planning, and no one else is involved.

2. The farmer takes the tomatoes to a village and sells them to a trader who has a stall. The owner of the stall inspects the tomatoes (looking at size, colour, surface marks, etc.), decides what price to pay the farmer, and hopes to get back in sales enough money to cover the outlay and make some profit.

3. The trader has contacts in a nearby town or city, perhaps a friend who is a wholesaler. The wholesaler purchases a lorry-load of tomatoes each week, probably from several traders, and sells them to retailers in the town who regularly deal with him. They negotiate a price between them. Alternatively in a large city the wholesaler may sell his lorry load of tomatoes at an auction, as at Kariakoo Market in Dar es Salaam.

4. Another trader has an arrangement to supply a processing factory making tomato puree or tinned tomatoes. The factory needs a regular supply of good quality tomatoes, at the lowest possible price. The trader agrees to deliver a lorry-load of tomatoes each week during the period of the tomato harvest.

5. This trader signs a contract with several farmers before they plant their tomatoes, guaranteeing to purchase their harvest at an agreed price. The contract specifies the quantity and quality of tomatoes required, and the price. The trader may supply inputs to the farmers on credit, deducting the cost from the farmers' payments. This arrangement, known as "contract farming", is considered in more detail in the next chapter.

6. After purchasing tomatoes from farmers or traders, a factory may pass its product on to another factory, which makes a more complex product. For example the tomato paste made by the first factory may be used in pizzas made by the second factory. These will be sold to shops and restaurants. The pizzas are likely to include cheese (also an agricultural product), and a variety of vegetables and spices, and its base is made from wheat flour. A single pizza may thus include ten or more agricultural products.

MARKETING FUNCTIONS AND VALUE CHAIN ANALYSIS

As explained above, the post-harvest system encompasses a diversity of activities, including processing, distribution and marketing. These span the continuum from the field to the plate. Major technical concerns of the sector include the prevention of food losses and the preservation of food quality through proper handling, quality control and storage. A primary objective of the sector is to increase the "value added" in order to encourage farmers to increase the value of their production – if the "downstream" services, including processing, marketing and transport, are more efficient, then in principle more money can be paid to the farmers.

Marketing functions include exchange (buying and selling), physical activities (assembly, transportation, storage, processing, packaging, grading and standardisation), and facilitating activities (financing, risk-bearing and gathering market information).

Each of the ways of marketing tomatoes described above has a "value chain" – very simple at the top of the list, complex near the bottom. The value chain tracks the prices paid for a crop as it moves from farmer to trader, wholesaler, processor, and eventually to a consumer. It shows how value is added at each stage of the chain.

It starts with a final use and the price of the product in that use. The costs incurred up to that point are then calculated or estimated - storage, transport, packaging, and a tiny part of the costs of running the retail outlet where the end product is sold. If the price paid for the product at each stage in the value chain is known, the profit taken off at each stage can be calculated. When all the marketing and processing costs are deducted from the value of the crop in its final use, it can be seen how, if other players in the value chain had been paid less, farmers could have been paid more.

If a single agricultural product is involved, it is fairly clear how the value in its final use is shared among the different people who handle it on its path from farm to consumer. If the product is processed into a complicated manufactured product, such as a pizza, there is a value chain for each one of its components. This is likely to show that only a small proportion of the price paid by a consumer goes to the farmer. Most of the value goes to those who carry out tasks near the top of the value chain.

In practice few individual farmers, or even co-operative societies or unions, will have the capacity to undertake a full analysis of a value chain, even for a simple case like maize or rice in Tanzania. To do so requires a specialist who collects and compiles the data. If this has been done once, it will be relatively easy to update a second time, say a year later. The way the method works can be seen in the articles [1], [2] and [3] in the Further Reading for this chapter. But prices and costs vary, often from week to week, so any such analysis is only a guide as to who is getting the most benefit from the crop. It is useful more as a tool to assist farmers who want to argue that they should be paid higher prices than as a method that can be easily applied to show a farmer what he or she should be paid on a particular day.

MARKET IMPROVEMENT

Farmers often report significant losses during and after harvesting. Losses of as much as 50% at various stages in the value chain have been reported (see [4]). These losses are usually caused by pests and diseases, but often they occur as the result of deterioration through careless handling and inadequate storage.

Good quality storage facilities are needed, especially for crops that are perishable and those that are not resistant to post-harvest pests (this is a problem with some of the new high-yielding varieties). Improvements in transportation can reduce the time taken to bring crops to the market, and this can reduce crop losses and speed up payments to farmers. Reliable deliveries also make it possible for processors and wholesalers to operate with less stock. Such improvements can often make a bigger difference to farm incomes than changes in the method of cultivation or the use of fertilisers or other chemicals.

For many perishable products, such as fruits, tomatoes, cassava and possibly cashew nuts, the long-term challenge is to help farmers to make the transition from a low-yield agriculture, where the crop is sold in local markets, to a higher yield, higher volume production using agricultural inputs and better technologies, and supplying a processing factory which needs large quantities of produce. However such a factory may only be able to make a profit by paying the farmers less than they can get on local markets (where prices may be reasonably high, though the amounts that can be sold are limited). For this reason, attempts in Africa to construct factories for canning fruits, making tomato paste, or cassava flour have almost always failed, and will continue to fail unless the marketing function is improved and farmers see that they can make more profit by selling a large quantity to a factory at a low price, rather than a small quantity on a local market at a higher price.

INSTITUTIONS AND ORGANISATIONS IN THE MARKETING EFFORT

This section is devoted to an assessment of the strengths and weakness of the institutions and organisations involved in agricultural marketing.

Private sector traders

The marketing of agricultural products is mainly in the hands of the private sector. The amounts traded are often small – what the farmer can carry to the market. The farmer may sell directly to a consumer, and receive the full price. More commonly, even when the quantities are very small, the farmer sells to a trader, who assembles the produce until there is sufficient quantity to arrange transport to a city (see, for example, Wegerif [5] for the tiny margins on which many such traders operate).

As explained above, wholesalers may buy from markets where the crop is assembled, from traders, or from auctions in major cities. Wholesalers sell to sub-wholesalers, retail shop keepers, stall holders, street vendors, hawkers, hotels and restaurants – and in some cases to processing factories. Retailers are in direct contact with consumers and are well aware of their needs. The complexity of the marketing effort sometimes also requires the presence of facilitating agents, such as brokers, auctioneers and financiers.

Farmer groups

But the private sector can be a fickle friend. There can be a high risk of small farmers being cheated by unscrupulous traders, especially when there is only a single trader willing to make the journey to buy their produce, so that farmers are at the mercy of this trader and have no option but to accept the price he offers. Even when there are several traders they may collude to set low prices that exploit the farmers – cartels of produce purchasers are notoriously common.

To counter this risk of exploitation, and to gain the advantages of higher prices when larger quantities of a crop are sold, small farmers have long been encouraged to work together in groups for marketing purposes. The number of farmers in a group may be quite small – perhaps twenty – so that they all know and trust each other. There still needs to be some administration – to reject produce of poor quality, and to ensure that when farmers are paid, the money is divided correctly among the group – but the management needs of a farmer group are quite simple.

Primary co-operative societies and co-operative unions

On a larger scale, farmers can work together to market their produce through primary co-operative societies and co-operative unions. The co-operative societies around Lake Victoria, which grew rapidly in the 1950s, were initially "independent weighing posts" where farmers could check the weight of the cotton they were selling, and ensure that they were not being cheated by private traders using false weighing scales. Later the co-operatives became more formal farmers' organisations regulated by government.

A primary co-operative society (PCS) is a voluntary group of farmers. These societies are registered under an Act of Parliament which requires them to have a constitution, to elect officers, and to appoint staff – all following open procedures. The Act also requires that their accounts be audited, and that the societies are regulated, to ensure that they are using resources as agreed, and keeping their premises and equipment in good order. A PCS that purchases crops from its members borrows money on a short-term basis to pay the farmers, and then repays the loan when it sells what it has purchased – either to a private trader, or to a public sector body such as a Marketing Board, or at an auction. It may also borrow money on behalf of its members, to purchase seeds, chemicals or equipment. These costs are deducted several months later from the payments to the farmers when the PCS sells their crops.

In some cases the government passes legislation that requires farmers who grow a crop to be members of such a PCS, and only to sell their crop through that society. An example of this is the 2001 Tobacco Industry Act in Tanzania. The risk here is that farmers may have little trust in the PCS, if they are obliged to belong to it rather than choosing to join of their own free will.

It is one thing to have a PCS and quite another to have an effective one. The societies are likely to work efficiently only if they can provide better services in competition with private traders. The viability of a PCS may be undermined by mismanagement, embezzlement or misallocation of funds, leading to the demise of the society. In such cases smaller informal groups of farmers who know and trust each other may be preferred.

Co-operative Unions are apex organisations linking a number of primary societies. In addition to their role in marketing, these may get involved in primary processing, for example of rice or sunflowers, or in running maize mills, cotton ginneries, coffee pulperies, cashew nut-shelling factories, or other economic enterprises such as hotels and garages. But the history of co-operative unions in Africa has not been encouraging – many of their leaders became involved in corrupt practices, and their capacity to run and make profitable the enterprises they started was often limited (for more on this, see [6]).

The co-operative movement has thus had a chequered history in East Africa. At various times, when a PCS or Co-operative Union has failed economically, or when corrupt practices have been revealed, governments have imposed external management or even abolished the co-operatives (as in Tanzania in 1976), but they have always been recreated in similar form since farmers need an effective organisation for marketing their crops and for other purposes. See the Tanzania government report [7] on the steps taken to keep the co-operative movement alive, the paper by Biddy [8] and the Case Study for this chapter which is an extract from that study.

The public sector – marketing boards

In the colonial period, and in the years afterwards, the state played an important role in the marketing of key export crops. It set up marketing boards – which were often the only bodies permitted to purchase and sell a crop. This gave them a direct interest in the quality of a crop, and they employed inspectors to ensure, as far as possible, that approved practices were adhered to. Often this depended on legislation which made it illegal to carry out certain practices (such as packing cotton in bags made of polythene – which may reduce the quality of the cotton), or required farmers to carry out certain tasks (such as burning cotton plants at the end of the season, to control plant diseases). They employed specialists who understood overseas markets, and knew what was needed to get the best prices. They inspected the accounts, and the state of factories and stores and the machines and equipment of the factories which carried out primary processing.

Many purchasers had personal contact with officials of the marketing boards, and they paid premium prices for reliable and good quality produce. This was not the case however when marketing boards sold at low prices to the colonial government. A devastating study by Robert Bates [14], drawing most of its evidence from West Africa (though some also from East Africa), showed how marketing boards were used by colonial governments to tax farmers, and how these bad practices continued after Independence.

In the 1960s and 1970s in Tanzania most of the marketing boards operated with "pan-territorial prices", i.e. at the start of each season they announced the price that farmers would get for a crop, and it would be the same anywhere in the country. The marketing board met the costs of transport and storage and carried the risk of losses if sale prices fell. One consequence of this was the promotion of increased production in remote places where previously the crop might not have been profitable because of small quantities and high transport costs. A recent study by Wuyts and Gray [9] interprets what happened as follows:

> ... looking from hindsight, pan-territorial pricing should actually be seen as an example of a successful *infant industry* policy. The reason is that its effects in creating a new spatial structure of food production capabilities were sustained well beyond the demise of the actual policy initiative that gave rise to them. ... pan-territorial pricing radically changed the regional distribution of marketed food production in Tanzania.

However, the marketing boards found it difficult to set the prices at the correct level. If they set them too high, they might have to dispose of the crop at a loss. If too low, farmers would sell their crops elsewhere. Either way they could easily lose money. That is one reason why most of the marketing boards that exist now no longer purchase crops. They are regulators only, although often without the staff or expertise to do even this effectively.[10]

FIRST AND SECOND STAGE PROCESSORS

Most crops require some kind of basic treatment before sale, for example sunflower seeds need to be dried in the sun, cashew nuts to be removed from the "apples" on which they grow. The basic aim of processing is preservation, but associated with this is improvement of the produce to increase its value, and also the development of new consumer goods. Elementary forms of processing to preserve agricultural products are widespread in all societies, for example the sun-drying of sweet potatoes by the Sukuma people in Tanzania, and meat or fish preservation through smoking by many ethnic groups in Africa.

---

[10] A few marketing boards do still purchase crops: in Tanzania the National Food Reserve Agency buys maize and small quantities of sorghum when there is surplus production, and the Cereals and Other Products Board has can purchase food crops which might otherwise be wasted.

There are, typically, at least two stages of produce processing. **Primary processing** involves cleaning the crop and removing waste material. Examples are the groundnut decortication, paddy husking, maize milling to make flour, sunflower pressing to extract oil, cotton ginning to separate lint from seed, coffee pulping to remove the beans from the cherry, cashew nut processing to separate the nut from its shell. Some of these primary activities take place away from the farms where the crops were grown. For example sisal factories separate the fibres from the green matter around them; tea factories cut and dry the tea leaves to preserve them; dairies pasteurise milk and perhaps make other dairy products such as butter and cheese; abattoirs produce meat and hides, and sugar mills process sugarcane. These activities help to reduce the weight and volume of produce – making it more easily transportable. They also enable farmers to extract what is useable, add value to it and further contribute to storability.

The next step – **industrial processing** – is the manufacture of finished consumer goods ready for use – for example making instant coffee from coffee beans, sausages and burgers from meat, bread and cakes and other baked products from flour, brewing beer from cereals such as barley, producing canned fruits and vegetable products, or packaging them in some other way, manufacturing cigarettes from tobacco, textiles and garments from cotton, shoes or other leather items from hides.

Some of these processing factories are large and costly. Others operate on a small scale to supply local markets. The great majority are owned and run by the private sector. The owners of a factory may also run an estate which produces the crop. At the other extreme the owner of a small mill may wait for consumers to turn up with small quantities of rice or maize, and mill them for a pro rata fee. Factories owned by large companies and supplying national markets employ marketing specialists, and spend millions on advertising on TV and radio.

Regulators, such as auditors and inspectors

The purchasing and processing of crops raises many issues: the prices paid to the farmers, the safety and quality of the processes and of the resulting products, and the payment of local and national taxes.

Inspectors and regulators are usually government employees who attempt to control abuses. The earliest inspectors checked weights and measures to ensure that farmers were not being underpaid, and their successors continue this work today, when they can. Factory inspectors check the safety of machines and factory buildings. Health inspectors check premises for food safety, and examine products to ensure that they do not contain dangerous substances. Inspectors from marketing boards or crop authorities check the quality of products and the manner in which they are processed. Tax inspectors check the account books to ensure that taxes are paid. Auditors attempt to identify fraud. Most inspectorates are backed up by laws which give them powers to gain entry to premises, to inspect account books, to interview key staff, to issue warnings or fines, to take offenders to court, or even to close down businesses or premises completely.

However, taking a business to court is slow and time-consuming, and there is a loss of employment if a business is forced to close. So most inspectors and regulators prefer to warn those they discover breaking the rules, and prosecute only if the offenders fail to improve. Regulators themselves need to be regulated – they are very vulnerable to corruption from those they are regulating.

In most African countries, there are few inspectors and many failings – and only a few of these are dealt with. The main exception is for crops or livestock products exported to the European Union or other countries which have very stringent requirements regarding any crops entering their country. This, and other accreditations, are discussed below.

Accreditation for premium prices

It is possible for a farm, or a group of farms, to gain an accreditation from an outside body for its agricultural practices and products. Farms with accreditation can use a trademark on their packaging. The accreditation guarantees that the product is of high quality, or that the farmers are not destroying the environment, or that those who produce the product are not being exploited. These guarantees often motivate purchasers to pay higher prices, particularly in some overseas markets in Europe and America.

Many countries have very high standards which agricultural products being imported have to meet. Products are inspected at ports of entry, and if they fail the tests they are seized and destroyed. Regulations may cover the quality and safety of a crop, regardless of how it is produced. Examples of these are the product's "shelf life" – how long it will keep its quality, and its appearance – and its freedom from contamination, including a lack of pesticide residues, or bacterial contamination. Some of these regulations are more about the appearance of a product than about its quality (for example oranges with green skins are not acceptable for sale in many European markets).

Accreditation schemes were set up by the large supermarket companies in Europe to ensure that their suppliers met their quality conditions. An accredited product is unlikely to fail the tests at its port of entry (unless there were delays in the export process, so that the product had started to deteriorate), but buyers may want standardised sizes and qualities – apples or tomatoes of similar sizes, for example, which can be guaranteed to stay fresh till a given date. Produce which is blemished, or stained, or of irregular shapes, may be refused by buyers, or get very low prices.

Other schemes recognise farmers who are following "fair trade" practices: paying fair wages, ensuring that working conditions are safe, and that practices will not contribute to climate change. Some schemes go much further, and recognise *organic* farming, that is crops that are produced without the use of inorganic chemicals. A case study of one fair trade accreditation scheme involving tea growers in Tanzania and Uganda is included at the end of this chapter. All such schemes require the farmers to keep clear records of what they do on their plots, including the inputs they use. Independent inspectors are employed to visit the farms and check that the practices are being observed. The farmers have to pay a fee to cover their costs,

The main accreditation schemes in East Africa are:

GlobalG.A.P. (Good Agricultural Practice) [10]. This is the largest farm produce certification scheme in the world. It was set up to create common standards for supermarkets and other purchasers of agricultural products, at first in a small number of countries, but now around the world. It certifies the quality of the crop and its safety, creates an audit trail that traces where the crop was grown, as well as checking on workers' health, welfare and safety, environmental sustainability of the farming practice, and the welfare of any animals involved. The scheme operates in Kenya and Ethiopia, and to a smaller scale in Tanzania, where it certifies exports of beans, avocados and peas.

The **BRC** (British Retail Consortium). The BRC has produced a "Certified Suppliers Guide" – a global standards directory with details of BRC-certified food, packaging and consumer product suppliers, guaranteeing a minimum standard. The BRC agent in Kenya is SGS, based in Mombasa, which also works in Uganda, Ethiopia, Rwanda, South Sudan and Somalia. It is similar to the GlobalG.A.P. scheme.

Fair Trade: The **International Fair Trade Certification Mark** emphasises the employment conditions of workers, to guarantee that they are paid what is regarded as a "fair" wage – above minimum wage levels in most African countries – and are employed with good terms and conditions. Products with these accreditations can expect to receive higher prices.

Traidcraft Exchange and **Traidcraft PLC**. Traidcraft Exchange works with groups of farmers, to ensure that their products are produced using sustainable agricultural practices and that they pay fair wages to any workers they employ, or, where possible, who are employed by their suppliers on other parts of the value chain. Traidcraft PLC is a trading company in the UK which purchases from "ethical" companies or farmers' groups such as those assisted by Traidcraft Exchange. Its products carry the Fairtrade logo, and are marketed through supermarkets or agents.

The East African Organic Products Standard [11] ("Kilimo Hai" in Tanzania [12]). The organic farming movement has strict standards. The East African Organic Products Standard has a number of requirements: chemical products which may cause harm to the environment may not have been used on fields for at least a year before registration is accepted; as far as possible naturally occurring products must be used; Genetically Modified Organisms must not be used; biodiversity should be encouraged; monocropping must be avoided; farmers must ensure that livestock have sufficient space to move around freely, and, if purchased feed is used, it must be grown organically. Synthetic chemicals and drugs may be used only if there is no alternative. There are bodies in most countries which promote organic farming, for example the Soil Association in the UK and Sustainable Agriculture Tanzania.

There are, however, paradoxes. One is that a great deal of farming in Africa is organic by default, simply because the farmers cannot afford to purchase inputs. But it is not accredited as such, and so the farmers do not get premium prices.

Another paradox is that organic and fair trade accreditation does not reject farms that use large-scale mechanisation. So farms may employ very little labour, and may contribute to global warming through the use of fossil fuels to power machinery, but they are still accredited.

A third paradox is that many systems of accreditation for organic products permit naturally occurring minerals to be used. These include guano. This, however, is a product of the faeces of bats or birds. Much of it was laid down hundreds or thousands of years ago, and it is being used much faster than it is put down. In that sense it is not a renewable source. These matters are further explored in Chapter 10.

MARKET INFORMATION SYSTEMS AND MOBILE PHONES

Many marketing opportunities for small farmers are lost because of the lack of information on price levels at different points in the marketing chain or in different places. Thus there may be a shortage of bananas in a particular market – perhaps a lorry has broken down – but traders or farmers who could benefit from this do not get the information in good time. This is particularly a problem with perishable products, such as many vegetables or fruits.

"Market access" projects work with farmers to identify the problems they face in marketing their produce, and to find solutions to alleviate the constraints. Market information systems are designed to collect, process and disseminate this information. They can be managed by public or parastatal institutions, professional organisations, NGOs or donor-funded projects.

In the short term, the type of information farmers need may include:

- Who and where the buyers are; how they can be contacted; their conditions of business; their preferences for varieties, packaging and delivery
- When the markets are available and how to access them
- Current prices and information about surpluses or shortages
- Longer term projections of future prices.

Such information can help farmers to:

- Decide where to sell their crops
- Check on the prices they are getting
- Decide whether or not to delay selling (if they can store their produce safely)
- Decide whether to grow out of season produce
- Decide whether to grow a different range or variety of crops

There have been many attempts across Africa to establish market information systems based on mobile phones. They have not all been successful, and sometimes they benefit traders more than they benefit farmers. The information has to be compiled. It can then be put in a text message which is sent to a list of subscribers. Or it can be accessed on demand – for example as a page on Facebook or WhatsApp. But it is often difficult to get enough subscribers who are willing to pay for this service, to make it a commercial proposition.

MVIWATA is an organisation which supports groups of farmers in Tanzania, described in reference [13]. It has a project to support local markets, managed by boards of local stakeholders. Information about current prices for produce is posted in the market centre and in surrounding villages. The market becomes an active information and commercial centre which gradually encourages farmers to make use of its service. An SMS-based platform developed by MVIWATA in 2012 connects its members to a broader range of markets and traders around the country. Staff are paid by local councils – nominally out of extra taxes collected from increased quantities of produce passing through the markets. Whether this model will survive the end of subsidies from external donors remains to be seen.

A market report on a local radio station, in which a reporter goes to the market and reports on key prices and price changes, may at present be more practical than a service run by salaried staff. Those interested can tune in at a regular time. But the amount of information that can be conveyed, and its reliability, will be questionable.

Meanwhile, the biggest innovation using mobile phones in marketing is almost certainly **M-Pesa**, which makes money transfers possible without the need for large quantities of cash. M-Pesa has greatly reduced the risks of robbery and fraud.

CONCLUSIONS

This chapter has considered many aspects of crop marketing, and there is further discussion of marketing in the chapter which follows.

Marketing is key to the success of agriculture in Africa, as elsewhere. For when markets do not exist, or are poorly managed so that prices are low, the farmers are unlikely to take production seriously. On the other hand, if marketing is efficient and prices are good, production is likely to increase.

The first part of this chapter looked at how the concept of a Value Chain can indicate where marketing is inefficient, and where it can be improved. If markets are more efficient, then, in principle, farmers can be paid more.

The chapter then looked at the main institutions involved in agricultural marketing – in the public sector as well as the private. This was followed by discussion of accreditation schemes for organic or fairly traded produce, and of the use of mobile phones to provide information about prices that will assist farmers.

Here are some of the main points:

1. Much more attention, and investment, is needed in marketing. Extension workers, NGOs and governments work hard to increase primary production. But if the marketing systems are not efficient and reliable, there is little point in increasing production.

2. Analysis of value chains shows that often the best way to get more value to farmers is to invest money off the farm, through improvements in storage, transport infrastructure such as feeder roads or bridges, better packaging, and marketing.

3. When the quantities produced by individual farms are small, effective marketing depends on co-operation between growers, through producer groups and/or primary co-operative societies and co-operative unions. But experience of these has at best been mixed: they are prone to mismanagement, so they need constant oversight by members and regulating institutions.

4. Marketing is open to corruption. It therefore depends on effective regulators and inspectors, who will prosecute those found to be cheating the farmers, and active police and justice systems. This applies to co-operatives as much as to private traders.

5. Some aspects of corruption may be controlled by competition, for example by having many traders. But even then the traders may collude together to cheat farmers.

6. Accreditation schemes can result in higher prices for farmers, but only when consumers are willing to pay more for crops that are grown without the use of manufactured chemicals, with fair-wage labour, and without genetically modified seeds.

7. Marketing depends on good information – hence the value of systems which provide information to farmers about prices and availability of supplies and inputs – but it is not easy to find a means of financing market information systems that will be sustainable in the long term.

Case Study 7: Tanzania's Co-operatives Look to the Future

This case study is an edited extract from a report by Andrew Bibby [9], funded by the International Labour Organization as a contribution towards the implementation work of Tanzania's Co-operative Reform and Modernisation Program.

This is a time of change for Tanzania's co-operatives.

Co-operatives in Tanzania have a long history, dating back to the late 1920s. In times past, they played a vital role in the rural and urban economic and social development of the country.

However, more recently the image has become a negative one. For many people in Tanzania, co-operatives are seen as stuck in the past, unable to cope with modern economic realities. Far from being models of member self-empowerment, their image is tarnished by poor administration and leadership, poor business practice, and by corruption.

Co-operatives developed historically – in Tanzania as elsewhere in the world – because they performed a valuable role. That role remains as relevant as ever today. Without co-operatives, small producers are left with almost no form of collective organisation, and thus at an immense disadvantage when taking their products or crops to the market. Without savings and credit co-operatives, many poor people have no safe home for their savings and nowhere to go for loans. (Co-operatives can also provide solutions through collective action in other areas, too, such as fisheries, forestry, minerals and housing.)

If they are to meet their potential in the future, a comprehensive transformation of co-operatives in Tanzania will be necessary. The task is to focus back on the key co-operative principle: that co-operatives are owned and controlled by their own members. The purpose of co-operatives is, above all, to fulfil their members' economic and social needs.

To achieve their goals, co-operatives need to be commercially viable enterprises, able to survive and prosper in the marketplace. To be sustainable, they have to be run on a businesslike footing. In contrast to other businesses, however, the rewards from their trading activities are available to be shared between all the members, on a collective basis.

In recent years, a number of important initiatives have put in place the foundations for a rebirth of Tanzania's co-operative sector. These include the 2000 Special Presidential Committee on reviving co-operatives and new co-operative legislation of 2003. These and other efforts have culminated in the production of a key strategic document, the Cooperative Reform and Modernisation Program 2005–2015.

This report (a 'home grown' initiative rather than the product of external consultants) has been produced with the direct involvement of the Tanzanian co-operative movement. It is honest about the problems: it identifies problems of poor management, inappropriate co-op structures, corruption and embezzlement, lack of working capital, lack of co-operative democracy and education, weakness of supporting institutions and, in general, an inability to compete in a liberalised market economy.

However, it also sets out detailed strategies for overcoming this problematic inheritance by creating:

- Economically strong co-operative societies, capable of facing competitive challenges.
- Strong savings and credit co-operatives, providing better services and offering a source of capital for co-operatives.
- An empowered membership.
- Good governance and accountability in co-operative societies.
- A network of co-operatives with efficient and cost-effective structures, able to respond easily to the needs of their members.

Promoting member empowerment and healthy co-operative democracy

Co-operatives are enterprises democratically owned and controlled by their members. To meet their full potential, Tanzania's co-operatives need to have members who feel able to participate and engage actively in the life of their organisation. As the Reform and Modernisation Program puts it, "it is only when the grassroots membership is empowered that Tanzania will see a true emergence of democratic and economically viable cooperatives". Thus far, it admits, the involvement and participation of members remains weak.

Strong emphasis is therefore given to the need for member empowerment. This is described as "a process of power sharing with the ordinary members, in order to build their confidence and their ability to manage their own economic affairs and their co-operative organisation".

Empowering members requires the creation of an enabling environment, one which encourages participatory ways of tackling and solving problems. It also requires a change of attitude on the part of co-operative leaders and support workers, so that they can become facilitators and agents for change.

Fortunately, there is already firm evidence that this can be achieved. Between 1996 and 2000, an innovative pilot programme promoting member empowerment took place in the Kilimanjaro and Arusha regions. A total of 171 primary societies participated in the project, known as MEMCOOP (Member Empowerment and Enterprise Development Programme). About 60,000 co-operative members went through a MEMCOOP training programme, and additional training was held for well over 2000 committee members and secretaries.

The programme successfully helped change the behaviour and attitudes of members, so that they felt a new sense of ownership over their co-operatives. There were strong business benefits, too. For example, coffee marketing co-ops participating in the programme obtained an average price of T.Shs.1,286 per kg of coffee in 2003/4; the average for co-operatives outside the programme was, by contrast, only T.Shs.600.

Establishing strong corporate governance

The Reform and Modernisation Program admits that co-operatives in Tanzania have, too often, suffered from poor leadership and bad management. Co-operative leadership has, it says, become associated with a lack of accountability of members, untrustworthiness and persistent corruption.

Good corporate governance goes hand-in-hand with effective member empowerment and healthy internal democracy. Moves to strengthen members' role in their own co-operatives also strengthens the co-operative's leadership and management.

Beyond this, a number of concrete actions are being taken to ensure that co-operatives are overseen by leaders who are honest and properly accountable. As a consequence of the Co-operative Societies Act of 2003, the Code of Conduct for co-operative management for the first time limits co-operative board members to no more than three three-year terms. After nine years, a board member automatically ceases to hold office.

There are also more rigorous procedures which are now in place for co-operative members wishing to stand for election to a board. The Code of Conduct provides for a minimum period of three years' active membership of a co-operative before a member becomes eligible for a leadership position. The Code also provides for minimum qualifications in terms of both co-operative education and formal schooling (normally co-operatives leaders are expected to have completed secondary education).

As a direct consequence of the Reform and Modernisation Program, a process is now under way whereby every primary and secondary co-operative in Tanzania will call a special general meeting called for the purpose of holding new elections to co-operatives' boards. These meetings and elections are overseen by the Registrar of Co-operatives or his appointed staff.

Candidates for leadership positions are required to apply in writing, giving full details of appropriate past experience. Board elections take place under the auspices of an Election Supervising Officer, acting with powers given to them by the Registrar of Co-operatives, and an election Panel, comprising at least four people "of recognised integrity" appointed by the Election Supervising Officer. The Officer and the Panel have the task of scrutinising applications for aspiring Board members and of rejecting those from candidates who possess insufficient experience and skills, or who have been previously associated with maladministration or malfeasance.

The Code of Conduct also introduces measures to help protect co-operatives from leaders who try to use their position to advance their personal interests. There are more rigorous requirements to be met when co-operative employees are to be appointed.

Transforming co-operatives into viable businesses

If they are to meet their wider social and community objectives, co-operatives need to be viable as commercial enterprises.

In some instances, co-operatives are still operating unprofitable facilities, such as hotels, oil mills or cotton ginneries. Without an emphasis on commercial viability, co-operatives will not be able to shake off their past poor image. As the Reform and Modernisation Program puts it, "Too many banks and other trade partners consider co-operatives as not creditworthy. They are simply bad business partners."

On the other hand, the Reform and Modernisation Program also recognises that, when the trade reform process was introduced in the 1990s, co-operatives were given very little support to help them adjust to a newly liberalised economy.

There is a need now to ensure that co-operatives can be helped to make the adaptations they need to run as sustainable business ventures. co-operatives are reminded that the competitive environment can be fast-changing, requiring a proactive entrepreneurial approach.

The Reform and Modernisation Program strongly encourages co-operatives to undertake strategic planning, and to draw up corporate strategies and action plans that can be used as tools in the day-to-day management of the businesses. Once drawn up, corporate strategic plans are also valuable as performance assessment tools for use by members in holding their management accountable.

The goal of creating a critical mass of viable and sustainable co-operative businesses in Tanzania requires co-operative leaders and members with improved business skills. A culture of entrepreneurship and business acumen has to be built.

Meeting co-operatives' need for capital

Tanzania's co-operatives need access to capital, to enable them to develop as effective, commercially viable businesses.

Historically, co-operatives attempted to build up their own reserves as a source of business capital. However, over time their capacity to build up internal capital became eroded, partly by government policies, and partly by poor business performance resulting from declining margins (for example, in the case of coffee, cotton and cashew nut prices). From the 1970s, co-operatives increasingly relied on government support for finance. This was given out as loans, but was very often not repaid. Co-operatives became trapped in a dependency/parasitic relationship with government which seriously weakened their ability to develop into sustainable business enterprises. This dependency also weakened member democracy.

Rebuilding Tanzania's co-operatives as effective member-owned businesses requires a clear break from this unfortunate historical legacy. Viable co-operative businesses today suffer from this legacy in a number of ways. Many commercial banks still view co-operatives as inherently not creditworthy, and co-operatives are often weighed down by the presence on their balance sheets of accumulated debts dating back many years.

What is already happening?

- The Tanzanian government, in approving the Co-operative Reform and Modernisation Program, agreed to write off T.Shs.23 billion of accumulated debts.
- Co-operatives, as well as other small businesses, are benefiting from the warehouse receipt system, currently being piloted in eight regions of Tanzania. This system, launched in 2004, is part of the Agricultural Marketing Systems Development Programme.
- A study of the possibility of a new National Co-operative Bank has been commissioned.

Conclusions

The objective of the Co-operative Reform and Modernisation Program is, in its own words, to initiate "a comprehensive transformation of co-operatives, to become organisations which are member owned and controlled, competitive, viable, sustainable and with capacity of fulfilling members' economic and social needs".

To achieve this, the program sets out both an analysis of past problems and a detailed route-map of the way forward. There is recognition that the renewal of the co-operative movement in Tanzania is a process, one which realistically will take ten years to achieve fully. However, there is also a clear understanding of the need for immediate action.

The goal of the current strategy is not simply to help existing members of Tanzania's co-operatives to revive their organisation. Success depends also on reaching out, to ensure that new members are attracted and that, where economically viable, new co-operatives are established. The Co-operative Societies' Rules of 2004 provide a detailed framework for what are called 'pre-cooperative groups', which can easily be set up as a first step towards formal co-operative status.

There is also great scope for Tanzania's Community Savings Organisations (SACCOs) to develop, so that a much larger percentage of the population has access to an easy and trustable place to go for saving and borrowing. There may be opportunities, too, to encourage the creation of co-operatives in other sectors. Agricultural co-operatives, important as they are, are only one area of the economy where collective power can be of benefit.

Case Study 8: Cafédirect – Fair-Traded Tea from Tanzania and Uganda

This case study is based on a review of Cafédirect's partnership with tea growers in four locations in Tanzania and Uganda, undertaken by Antony Ellman over the period 1999–2002. While the current absolute production and cost figures have clearly changed, the relative figures are thought to be broadly comparable to those in the review period.

Cafédirect is a fair trade company in the UK with a subsidiary brand called Teadirect. Purchases of coffee form the bulk of its business, but Cafédirect also buys tea on fair trade terms from smallholders and commercial plantations in Tanzania, Kenya and Uganda. This case study describes the terms on which the tea is bought, and outlines the benefits to the producers, the companies and the ultimate consumers.

Cafédirect buys made tea from four main producers with profiles as indicated in Table 8.1. Each supplier has both a core plantation managed by the factory company (or by a co-operative society in the case of the Ugandan producers) and a system of smallholder outgrowers from whom green leaf is bought for processing and marketing by the factory.

Table 8.1: Profiles of Four Teadirect Suppliers

Suppliers	Kibena Estates, Tanzania	Herkulu Estate, Tanzania	Kayonza Tea Growers Co-operative Society, Uganda	Igara Tea Growers Co-operative Society, Uganda
Area of tea (ha)				
Estate	690	230	50	120
Smallholders	800	400	2000	1800
Average annual tea production (tonnes)	2394	335	1432	1819
Average annual Teadirect purchases (tonnes)	124	17	55	
No. of workers:				
Factory	208	50	120	199
Field	875	95	100	178
No. of smallholders	1000	600	2500	1936

In buying tea from these suppliers Cafédirect follows its "Gold Standard" purchasing policy, which includes:

• Direct purchase from the tea producers
• A guaranteed minimum price not less than the current market price
• An additional "fair trade premium" for use in community activities
• Rigorous quality standards throughout the supply chain
• Transparent accounting procedures
• A partnership agreement between supplier and buyer
• Adherence to sustainable economic, social and environmental guidelines.

The percentage of each supplier's total production purchased by Cafédirect is quite small (ranging from 2.3% to 5.1%), reflecting the company's policy of spreading risks by avoiding over-dependence on any single supplier. However, the price offered for high quality tea is attractive, providing a strong incentive to producers to improve the quality of their tea.

Cafédirect has adopted a minimum pricing model for tea, which protects the producer when the market price is low, but also protects the buyer when the market price is high. The model takes the average price of sustainable tea production, independently assessed, adds an agreed profit margin for the producer (10–20% above the break-even price), and incorporates the "Fairtrade premium" set by the Fairtrade Labelling Organisation (FLO) for all fair-trade tea producers (in 2017 the premium was £0.40 per kg).

In addition Cafédirect provides an aid-funded Producer Support and Development Programme, which aims to help producers to improve the efficiency of their production, to monitor the conditions of work of smallholders and employees, and to ensure that the benefits of tea sales are distributed equitably within both estate worker and smallholder communities.

Table 8.2 shows the average tea price received by the four suppliers during the review period, the margin over the average price at the Mombasa tea auction, the average annual receipts under the Fairtrade premium, and the annual expenditures made using this premium.

Under the terms of FLO registration as a fair trade partner, the tea producers elect a Fair Trade Committee representing the interests of all the stakeholders – smallholder tea growers, factory workers, factory and plantation management and community representatives. The Committees consider proposals received from any stakeholders for the allocation of Fair Trade Premium funds.

Table 8.2: Average annual receipts and expenditures from fair-trade tea sales

Suppliers	Kibena Estates, Tanzania	Herkulu Estate, Tanzania	Kayonza Tea Growers Co-operative Society, Uganda	Igara Tea Growers Co-operative Society, Uganda
Average Teadirect price ($/kg)	$2.11	$1.72	$2.09	$2.01
Average auction price for comparable teas ($/kg)	$1.74	$1.00	$1.77	$1.41
% margin of Teadirect price over auction price	21%	72%	18%	42%
Average annual Fairtrade Premium receipts in £ sterling (@£0.40/kg)	£49,876	£5,688	£34,605	£15,771
Average annual Fairtrade Premium use (in £ sterling)	£25,341	£5.516	£19,908	£14,963

Initially the Committees were helped with assessment of proposals through the Producer Support and Development Programme. Typical uses to which funds have been put include improvements to schools, health facilities, community centres, roads and water supplies, as well as more directly productive enterprises such as tea nurseries and infrastructure for facilitating leaf collection.

Although the scale of Fair Trade production and the sale of tea represents a small niche market, the experience described in this case study shows that fair trade can bring real benefits to both producers, buyers and consumers. The benefits realised include:

- Improvements in the quality of tea produced and in the efficiency of marketing
- Improvements to farmer and worker incomes and standards of living
- Greater trust and understanding between the producers and actors further along the supply chain
- Adoption of production methods which are more environmentally and economically sustainable
- Opening up of new markets for higher value commodities.

Such trading relationships merit further extension and replication as far as market conditions permit.

Further reading on open access on the internet

1. C. Martin Webber and Patrick Labaste. Building competitiveness in Africa's agriculture: A guide to value chain concepts and applications. World Bank, 2009. https://openknowledge.worldbank.org/bitstream/handle/10986/2401/524610PUB0AFR0101Official0Use0Only1.pdf?sequence=1

2. Lilian Kirimi, et al. A farm gate to consumer value chain analysis of Kenya's maize marketing system. Tegemeo Institute of Agricultural Policy Development. Egerton University. WPS 44/2011, 2011. http://ageconsearch.umn.edu/bitstream/202597/2/WP44-A-Farm-Gate-to-Consumer-Value-Chain-Analysis-of-Kenya-M.pdf

3. Ntemi Nkonya and Alethia Cameron Analysis of price incentives for cashew nuts in the United Republic of Tanzania 2005–2013. FAO Technical Notes Series, 2015. http://www.fao.org/3/a-i5041e.pdf

4. Alliance for a Green Revolution in Africa (AGRA). 2014. Establishing the status of post harvest losses and storage for major staple crops in eleven African countries (Phase II). AGRA: Nairobi, Kenya. http://agra.org/test/wp-content/uploads/2016/04/establishing-the-status -of-postharvest-losses-and-storage-for-major-staple-crops-2014.pdf

5. Marc Wegerif *Feeding Dar es Salaam: a symbiotic food system perspective* PhD thesis, Wageningen, 2017. http://edepot.wur.nl/414390

6. Sam Maghimbi. *Co-operatives in Tanzania Mainland: Revival and Growth*. ILO CoopAfrica Working Paper No. 14, 2010. http://www.ilo.org/public/english/employment/ent/coop/africa/download/wpno14co-operativesintanzania.pdf

7. *Co-operative Reform and Modernization Program*, United Republic of Tanzania 2005 http://www.tanzania.go.tz/egov_uploads/documents/co-operative_reform_and_modernization_program_crmp_sw.pdf

8. Andrew Bibby. Tanzania's *co-operatives look to the future*, 2006 http://www.andrewbibby.com/pdf/Tanzania.pdf

9. Marc Wuyts and Hazel Gray *Situating Social Policy in Economic Transformation: A Conceptual Framework*. ESRF Discussion Paper No.66, 2016. http://esrf.or.tz/docs/THDR2017BP-4.pdf

10. GlobalGAP. Tourstop 2016 Presentation, by Christi Venter https://www.globalgap.org/export/sites/default/.content/.galleries/Pictures/TOUR2016/TOUR2016_Ethiopia_Presentations/Introduction-to-GLOBALG.A.P.-Christi-Venter.pdf

11. *East African Organic Products Standard*. East African Community, 2007. http://www.ifoam.bio/sites/default/files/page/files/east_african_organic_products_standard_english.pdf

12. Kilimo Hai: Going Organic in East Africa. Video. International Federation of Organic Agriculture Movements (IFOAM), 2013. https://www.youtube.com/watch?v=Nkmk11IKwfI

13. Maggie Okore *Agricultural Markets and Poverty Reduction – the case of MVIWATA in Tanzania*. MVIWATA and AGRA, 2014. http://www.mviwata.org/wp-content/uploads/2014/09/Documentation-of-MVIWATA-Rural-Markerts-Experience-by-AGRA_Final.pdf

Reference not on open access on the internet

14. Robert Bates *Markets and States in Tropical Africa*, University of California Press,1981

Topics for Essays or Exam Questions

1. In many African countries internal markets are not well developed. What constrains or prevents the development of these local markets?

2. How can an efficient marketing system contribute to greater consumer satisfaction?

3. For some agricultural products a few buyers of a crop seem to dominate. Give reasons why that would be so.

4. Is there a risk that large food-processing companies may find it easier to import their raw materials instead of buying locally?

5. One of the problems in operating a processing plant is that of finding reliable supplies of raw materials. How might this problem be solved?

6. Should co-operative marketing be encouraged? Do you think that co-operatives can develop without government support?

7. Read the extract 'Promoting member empowerment and healthy co-operative democracy' from Andrew Bibby's report in Case Study 7 above. Its main conclusion is that "co-operatives need to have members who feel able to participate and actively engage in the life of their organisation". Is this realistic? How is it possible for members to protect their interests and prevent abuse and fraud?

8. Why are inspectors or regulators important in systems of agricultural marketing? Compare the tasks carried out by any two regulators, for example auditors of co-operative societies and inspectors employed by the Warehouse Receipt Licensing Board.

9. Discuss the relative merits of the main systems of accreditation of fair trade or organic products in Africa.

10. What part can fair trade play in giving higher prices and benefits to small farmers?

11. Explain how accreditation schemes can help farmers to obtain higher prices for the crops they sell. Consider two or three contrasting accreditation schemes in East Africa.

12. Why are accreditation schemes not more widely used by farmers in Africa?

13. Successful marketing depends on market information. Discuss the ways in which market information can be supplied.

CHAPTER 8

Resources for Agriculture
– Credit and Contracts

Key themes or concepts discussed in this chapter

- The role of the financial system in providing resources for agriculture.
- How banks underwrite the purchasing and marketing of crops.
- The risks facing banks or other financial institutions that lend to agriculture.
- The role of contracts in reducing the risks faced by farmers.
- The different types of contract farming, and the range of requirements that can be included in a contract to supply agricultural products.
- The strengths and the weaknesses of contracts, and why they can easily go wrong for either side.
- Warehouse receipt schemes, which are a means of linking storage and credit, and helping farmers get the best prices – but they depend on honest and disciplined staff, and have proved difficult to run in practice.

CREDIT AND CONTRACTS

The previous chapter laid out the fundamentals of agricultural marketing. This chapter looks in more depth at three aspects already mentioned. The first is *agricultural credit*. This is closely related to marketing because it is when the crop is sold that the farmer repays any loans. A similar situation applies to a trader or co-operative that borrows money to purchase a crop from a farmer.

The second more specialised form of marketing is *contract farming*, where farmers sign a contract to grow a specific crop and sell it to a specified outlet. The contract gives the farmer a guaranteed market. The purchaser may also supply the farmers with seeds and agricultural inputs on credit, deducting the costs of these when purchasing the crops.

The third is *warehouse receipt schemes* where farmers deliver their crop to a warehouse which stores it, in return for a "receipt" which states how much is deposited. These schemes are another means through which farmers hope to get higher prices for the crops they sell.

AGRICULTURAL CREDIT

Chapter 5 stressed that any kind of farming is a risky business.

Some farmers purchase inputs with cash, but most farmers cannot do this as they do not have savings. They can only obtain agricultural inputs if they receive them free of charge, or if they pay for them later, through some kind of credit arrangement.

Credit involves risks, for both the providers and the receivers of loans. For the farmers who receive loans, the risk is that they may be unable to repay the credit as agreed, because of a crop failure or because the price they receive for their crop falls below the expected level. For the lenders (eg a bank, trader or co-operative) they may be unable to recover the money they lent because the farmers choose to sell their produce to another buyer. This is called "side selling", where farmers sell outside the loan agreement in order to avoid deduction of the repayments that are due. It is one of the biggest problems facing agricultural credit in Africa.

A trader who has lent money to farmers also faces the risk that he may be unable to sell the crop that he has purchased – perhaps because of deterioration or poor quality of the product, or because of an unexpected glut in the market. If he has promised a processor, at home or abroad, a certain quantity and quality of a crop, and cannot supply it because the farmers did not deliver as expected, the trader is obliged to break the contract with the processor and will lose money and will be unable to get a further contract.

HOW BANKS WHICH LEND TO FARMERS GET THEIR MONEY BACK

Credit requires an agreement through which the receiver agrees to repay what is borrowed. For example a contract may specify that the borrower will sell produce to the buyer (a trader in the private sector or a co-operative). The buyer will deduct the costs of any inputs supplied to the farmer, plus any interest, before paying the farmer.

Where farmers or their families have documents ("titles") to the land they farm, which permit them to sell that land to someone else, a bank may "take a charge" on the land. Then, if the credit is not repaid, the bank can go to court and get approval to take the land away from that farmer, and sell it to get its money back. This works reasonably well for large farms in developed countries, but in Africa most small farmers do not have written titles. However, even for those who do have titles, banks are reluctant to go through the complex, time-consuming and often unpopular legal processes involved in taking land away from small farmers. So they are still reluctant to lend to these farmers on an individual basis. Instead credit agencies prefer to lend to farmer groups, or to producer co-operatives. This has the extra advantage that if one of the farmers in the group does not repay, the rest are still liable for the debts. The bank or credit agency gets its money back, and there is strong social pressure on everyone involved to honour the contract.

However, the overhead costs of administering small loan schemes are high, and banks and credit agencies have often found it hard to cover their costs and make money from these loans. For this reason many governments in Africa have concluded that the only way that the high yields associated with using fertilisers and other inputs may be realised is to give the inputs to farmers free of charge, or on a highly subsidised basis.

CROP FINANCE

Crop finance is money used by traders or co-operatives to purchase crops. This requires substantial quantities of cash, at the right time, usually just for the short period of time between buying the farmers' crops and selling what they have purchased to a wholesaler or processor.[11]

---

[11] The sums of cash held by traders can be reduced by the use of M-pesa. This is a system developed by mobile-phone companies which enables money to be transferred from one bank account to another by text message. Local agents hold reserves of cash, and pay out when requested. This method of paperless cash transfer is cheap and safe, provided the sums involved are not too great.

If basic checks are in place, bank lending to traders to enable them to purchase crops is a relatively safe form of credit, because a trader who does not repay will not be given further credit, and because the money is normally repaid within a few weeks.

FINANCE FOR FIXED AND MOVABLE ASSETS

Fixed assets are buildings, land, fixed processing machinery, and other machines or equipment which cannot easily be moved. Most fixed assets last for many years, so banks are more willing to lend against fixed assets than against movable ones. They may, on this basis, provide loans for part or all of the costs of purchase of entire farms.

Movable assets include tractors, lorries, agricultural machinery, or portable equipment such as pumps or small milling machines. Often these are supplied on a leasehold basis, in which case the lender (often associated with the manufacturer of the asset) remains the legal owner of the asset until the borrower pays off the full value of the asset, with interest. However, movable assets can get damaged in use, and their value depreciates extremely rapidly, so the lender will want to be paid quickly. The borrower will also have to insure the assets against damage or theft. Sometimes an asset disappears (though an expensive asset can be fitted with a GPS sensor which sends out a signal which will locate it).

Working capital. This was discussed in Chapter 6, where it was pointed out that there will be delays between investment on a farm and receiving money from sales of the resulting products. During that period wages have to be paid, vehicles have to be run and maintained, and there are also the costs related to individual crops, such as for seeds or fertilisers. This last may be met through credit secured against the crop and repaid when the crop is sold, as discussed above. Or it too may be met by a loan (or "overdraft facility") secured against the value of the land or of the company that owns it.

For loans to individual farmers, the issues of using fixed assets as security against repayment are similar to those of lending to other kinds of small business. An application for a loan should be supported by a robust business plan. The borrower should prepare a "risk analysis" of possible problems, including risks from possible price changes, poor weather, nearby competition, etc. The bank will undertake "due diligence", to check ownership and titles to the land, the assumptions of the business plan, and the track records of those involved.

When the risks are high, large investments may be more appropriately funded through equity finance, where the investor becomes a part-owner of the business by purchasing shares in a company which owns it, rather than through loans.

CONTRACT FARMING

Sometimes a farmer grows a crop on the basis of a contract with a company or a co-operative. This is called contract farming. The farmer agrees to grow the crop as specified in the contract and to maintain quality standards. The purchaser agrees to purchase it.

Many contracts include credit: the purchaser supplies seeds and other inputs to contracted farmers in advance, together with technical advice, and deducts the costs when the farmer sells the crop. The farmer gains a guaranteed market, at a price which may be higher or lower than would be achieved on an open market; the processor gets a guaranteed crop and can plan accordingly.

Interpreted this way, contract farming is not new and covers many situations, in most parts of the world, for farms large and small.

In Tanzania, for example, it featured in cotton growing around Lake Victoria, where ginnery companies were encouraged to contract with small-scale cotton producers to purchase their cotton, and to supply them with inputs such as sprays, with support from a local NGO, the Tanzania Gatsby Trust. The initiative, explained in more detail in the first case study at the end of this chapter, was undermined by side-selling and did not survive.

It has been more successful with tea production (the case study in the previous chapter), where the farmers have no alternative but to sell their green leaf to a nearby tea factory, with dairying (the second case study) where a sure market for a perishable product is needed, and with tobacco where under the terms of the Tanzania Tobacco Act contracted farmers are obliged to sell their tobacco to a Primary Co-operative Society, which enters into further contracts with a tobacco buyer and with a bank which provides credit for purchase of inputs and of the harvested crop (also described in the first case study).

It has also been successful with the introduction of a new crop, *Artemisia annua*, the source of an important drug for treating malaria, for which no market or production support would have been available if a processing company had not offered these facilities to farmers. This is the subject of the third case study at the end of this chapter.

CONTRACTS AND AGRIBUSINESS

Large agribusiness firms often use outgrowers. The underlying reason for this is that they do not have to pay wages for the labour, as they do when production is on their own estate or plantation. The costs of management may also be less. Furthermore, where land is in short supply, the use of small farmers may be the only way in which they can access a large volume of the products they require. They use contracts to ensure that quality standards are met, that inputs are available and used, and that produce is supplied as and when it is needed ([1, pp.1–2)].

In the last thirty years or so, two "revolutions" have brought contract farming to the fore. The first is the fast food revolution, led by two American companies, McDonald's and Kentucky Fried Chicken. These companies sell burgers and chicken, ready to be eaten, in industrial quantities. The second revolution, even more fundamental, comes from the supermarket chains which have developed and promoted the concept of the "ready-meal" prepared in a factory and then frozen or chilled, requiring only heating at home before serving. These changes raise many questions about a good diet, and have led to obesity as well as some loss of the skills of cooking. But their cheapness and convenience cannot be denied.

Supermarkets have also persuaded consumers that they can expect to purchase strawberries, apples, tomatoes, and other fruits and vegetables (and also cut flowers) at any time of year, sourcing them from different parts of the world. In 2014 the four largest supermarket chains in the UK accounted for over 54% of the food and non-alcoholic drinks sold there. When sales by smaller supermarket chains and on the internet were added, the figure rose to 90% – less than 10% was sold through small shops and in markets [2, p.13]. In the UK, purchases of fresh meat, vegetables, fruits, and flour from markets and small shops are declining.

The requirements for this kind of marketing of agricultural products – high standards of quality, certification to permit agricultural products to cross international borders, refrigerated and other specialised storage, and advanced logistics to track and order supplies – depend for their implementation on precise quality standards, timings, and quantities. It is difficult to achieve these without contracts, or, at the very least, less formal "gentlemen's agreements", with specific farms.

Sourcing by the international supermarkets is increasingly from a small group of countries, including Mexico, Chile, Spain, and Morocco.

Some supplies are from Kenya, Ghana and Côte d'Ivoire. Traditional auctions and commodity exchanges still exist, dealing with cotton, coffee, and non-perishable products such as cereals and oilseeds, but even with these crops contracts are often important (for example, as noted in Chapter 2, the brewer SAB Miller contracts with Mountainside Farms, a large farm in Tanzania, to purchase barley [7, Ch. 2]).

One of the downsides of these developments is the wastage of produce at every stage – in the fields, if a crop has been grown in excess of a contract; in storage or transport; at the factory, if the quantities contracted for turn out to be more than is needed; in supermarkets when products pass their sell-by dates.

WHAT CAN BE INCLUDED IN A CONTRACT?

Eaton and Shepherd [3, pp.58–9] set out how contracts may be specified. In some contracts, the only requirement is to sell to the contractor. At the other extreme, a contract may specify some or all of the following:

1. The duration of the contract – it could be for just one season, or for longer.
2. The way in which the price is to be calculated:
    *   prices fixed at the beginning of each season, perhaps including different prices for different grades of the product
    *   flexible prices, to be based on actual world or local market prices
    *   consignment prices, when the price that will be paid to the farmer is not known until the raw or processed product has been sold
    *   split pricing, when the farmer receives an agreed base price together with a final price when the sponsor has sold the product.
3. Quality standards required by the buyer.
4. A quota: usually a minimum quantity that the farmer will supply; sometimes a maximum.
5. Cultivation practices required by the sponsor, including the variety to be grown, inputs to be used, times of planting and harvesting, how the crop is to be planted, etc.
6. Arrangements for delivery of the crop.
7. Procedures for paying farmers and reclaiming credit advances.
8. Arrangements covering insurance.
9. Penalties for not fulfilling the terms of the contract – by either party – and sometimes rewards for successful completion of the contract.

The purchasing company may be involved in other ways. For instance it may also grow the crop on a farm it owns, and supplement its own production with purchases from outgrowers. This is the case in Tanzania with some sisal estates, the sugar plantations at Kilombero and Mtibwa, many tea companies, and the company Kilombero Plantations Limited which grows rice and is one of the flagships of the Southern Agricultural Growth Corridor of Tanzania.

The purchasers may be processors, such as the companies that operate ginneries in the cotton-growing areas of Tanzania, or mills that crush sunflower seeds to produce oil. Or they may be local traders linked with processors who use the contracts as a way of ensuring that they are able to purchase at least a minimum quantity of the crop. Or they may be major international or national traders in the commodity, as with the four companies that purchase the flue-cured tobacco grown in Tanzania. Or they may be farmers' groups or co-operatives, who themselves sign contracts to supply purchasers or processors of the crop. This has the advantage that there is an organisation to negotiate with the purchaser and to help farmers later if there are disputes. They can also help to ensure that loans are repaid as agreed.

As Minot [1] points out, there are other stakeholders. Farmers who are not in the contract may look with envy at those who are; on the other hand, they also have the freedom to plan their farms as they wish. Companies or co-operatives may wish to see contracts continue in order to sell specific products or seeds. Governments and local governments may see contracts as an easy way of collecting cesses or taxes. NGOs may assist by supporting farmers and strengthening their hand in negotiations with the purchasing companies.

ADVANTAGES AND DISADVANTAGES OF CONTRACT FARMING

Eaton and Shepherd [3] explore the possible advantages and disadvantages of contracts, for contractors and for farmers, showing that there are risks and potential benefits for both.

All long-term contracts depend on trust. When all is well, there should not be problems. But if some parts of the contract are not delivered, or if grievances –on one side or the other – are not dealt with, over time a contract can go sour. Trust, if lost, is hard to re-establish. In addition, the prices paid are often contentious. The farmers are in a position of weakness in relation to the contractors – except that they have the ultimate sanctions of sabotaging the contract by side-selling or ending the contract.

Contracts are seldom equally balanced. The processor, or marketing agent, has most of the power, but depends on the farmers to supply the raw materials, so has an interest in ensuring that they have the support they need. The farmers are comparatively weak; the purchaser knows that if it is not possible to agree a contract with one farmer, there will be others. Hence the case, made in the previous chapter, for farmers' co-operatives, or farmer groups, in which farmers work together to get the best deal for the group as a whole. This gives the farmers more control and bargaining power – though the reality may be different, as when co-operatives are corrupt or inefficient.

Contracts are commendable in principle. But in practice they depend on stable and reliable purchasers, processors and farmers, who value the stability of the contracts and who, in return, accept the conditions involved.

For farmers, the advantage of a contract is that it affords reliable and predictable incomes, and a guaranteed market for their produce. It also gives them access to credit, and they can benefit from extension advice provided by the contractor, and, in some situations, services such as ploughing or spraying.

But there are also possible disadvantages:

- A farmer may lose the possibility of selling the crop for a higher price elsewhere.
- There may be situations where the contracting company does not collect the crop as agreed, or is slow to pay for it, perhaps because at that time it finds that it can source the crop more cheaply on its central farm.
- The farmer may not want to be bound to a single supplier, or to be compelled to grow a minimum amount of the crop (with possible disruption to the growing of food crops for own use or local sales).
- The farmer may not want to accept the risks involved, if the crop should fail.
- Farmers may not trust the pricing regime and feel that they are being cheated.
- Farmers may not trust the quality requirements specified by the contractor, and how these relate to the prices they are paid.
- Farmers may not have confidence in the quality of the inputs supplied.

- Farmers may not want to depend on a single crop, with the risks and costs of controlling diseases and maintaining the fertility of the soil.
- Small farmers with contracts may feel that larger farmers are being favoured at their expense.

Many of the possible disadvantages for a contractor relate to these points:

- A farmer under contract may side-sell the crop for a higher price elsewhere, thus not supplying the quantities agreed, and avoiding paying back the credit.
- The contract may prove inflexible; for example if there are good harvests elsewhere, the contractor may wish to purchase the crop more cheaply from other growers.
- The costs of supervision and enforcement.
- In the long term the security of supply is uncertain.

The literature suggests that contracts are more likely to succeed if a crop is perishable, or if the quality declines if it is not sold quickly (as with tobacco, tea or sugar), or if it is expensive to transport long distances without processing (as with meat and dairy products, and many fruits and vegetables). Side-selling is hard in these situations.

Another situation where contracts can be useful is where a new crop is being introduced to an area, for which credit and a guaranteed market would not otherwise be available. A review of contract farming in Tanzania (Matchmaker Associates 2006) [4] has usefully reported and commented on contract farming for organic coffee, sugarcane, tea, tobacco, pyrethrum, sisal, chickpeas and milk, and on programmes then being developed for cashew nuts, fruits, paprika and other crops, and on marketing arrangements for many other crops where contracts are implicit but not written down. A study by Nakano and others [5] demonstrates how hard it is to introduce contract farming with a crop like rice which can easily be sold in local markets.

WAREHOUSE RECEIPT SCHEMES

Warehouse receipt schemes can take many forms, but what they all have in common is that farmers sell their crops to a strong and secure local warehouse, where they get a piece of paper, a *warehouse receipt*, instead of cash. These schemes have been promoted in many African countries by the FAO and the World Bank, as a means of enabling farmers to get the benefit of higher prices later in the season, and to get credit to purchase seeds and inputs for the following year's crop.

Here is how the Tanzania Warehouse Receipts Regulatory Board describes the system [6]:

The Warehouse Receipts System denotes a kind of trade by which commodities are stored in a Licensed Warehouse(s), the owner of the commodity receives Warehouse Receipts which certify the title of deposited commodities as of specific ownership, value, type, quantity and quality (grades). The Warehouse Receipt facilitates storage, future trade and access to credit, without necessarily moving the said commodities from the licensed warehouse. The Warehouse Receipts are therefore documents in hard or soft form issued in the warehouse by the Warehouse Operator, stating that the commodities certified in the Receipts are held in the warehouse and are at the disposal of the person named thereon.

The website from which this quotation comes has links to the extremely detailed legislation under which warehouse receipt schemes are set up. This includes no fewer than 32 forms and documents, many of them very detailed and requiring several signatures and supporting documentation for anyone running a warehouse. On pages 59–60 it reproduces a blank Warehouse Receipt form, as a farmer would receive it. The ambition of these schemes is almost overwhelming. But it is hard, reading the legislation, not to feel that all this detail will be very remote from most farmers.

There are many arguments in favour of warehouse receipt schemes [7] [8]:

1. The value of a crop may rise in the months before the next harvest. A farmer who has crops stored can sell them then, for the higher prices.

2. The value of the crop is not known until it is sold to a processor or exported. Thus the farmer may be paid a minimum price when he or she parts with the crop, and then a second extra payment later when the final price is known. In Tanzania this system is used for cashew nuts.

3. A warehouse receipt can be used as the basis for the supply of inputs. Thus instead of being paid in cash, a farmer may be paid in bags of fertiliser, or in help to buy a farm implement.

4. The produce in the warehouse can provide the security for other forms of credit. A financial institution may give credit equal to a specified percentage of the value of the stored commodity.

In some warehouse receipt schemes, the farmers have a choice. They can receive cash upfront on the day they sell their crop. Or they can take away a warehouse receipt. Or they can receive some cash, and a warehouse receipt for the rest. Then they can come back at any time, take more cash, and be issued with a revised receipt. Or they can take back some of their food stored in the warehouse, and exchange their receipt for an updated one.

In some warehouses, the produce of each farm is stored in sacks which are marked and kept separate. But the intention of those who invented these schemes was that the crops from all farmers would be mixed – and if the crop is stored in large silos this is the only way. Then a farmer who comes to take back some maize (say), will not get back the same maize that he or she sold. This means that very strict quality control must be applied, for if the maize from one farmer is of poor quality (for example, damp) it can affect all the crop stored; or if it is contaminated that will affect the value of the whole crop. It is therefore necessary that each sack be carefully inspected before acceptance, and that poor quality crop is rejected. Not always easy!

A warehouse can be in the public or in the private sector. Most are owned by Primary Co-operative Societies. But if these do not exist in an area, or are not trusted, the warehouse may be owned by a private trader or company. What matters is that the warehouse must always have sufficient cash to pay out as needed, and capable staff who can do the complicated calculations required in issuing the receipts. The operators must also be strong enough to refuse to purchase damp or poor quality produce, even when it is supplied by a powerful farmer, and not get sucked into corruption (such as forged receipts). They must keep very careful records and be open to the inspectors at any time. They must maintain the produce in good condition, so that the farmers and any banks who have lent money are certain that their loans are secure. They must be honest, because there are many ways in which a warehouse may cheat a farmer – through false scales, using incorrect prices, "mistakes" in calculations, thefts (covered up by fraudulent warehouse receipts, or incomplete records), or acceptance of poor quality produce.

The principle of warehouse receipts is good. But given these problems, it is not surprising that many farmers prefer cash in hand.

CONCLUSIONS

This chapter has shown that it is risky to lend to agriculture, and why many agricultural banks have not recovered the money they lent, or not sufficient to cover the costs of those who default. If this is to be avoided, the banks need to be fair but tough. They need to have clear procedures, backed up by contracts, to recover their money if loans are not repaid. They need to monitor carefully those they lend to, and borrowers need to talk to their banks if they encounter any unexpected problems. A bank will be wise to employ agricultural specialists to advise it, and to reduce its risks wherever possible. The advantages of lending to farmer groups or co-operatives rather than to individual small farmers are clear.

Contract farming to supply processing factories or supermarkets is also good in theory, but not always easy to sustain in practice. There are big problems if prices fall after a contract is signed. Or if some farmers manage to cheat the system by supplying low quality produce. Farmers need to understand the value of stable guaranteed markets, even if they sometimes get paid less than they could get outside the contract. Large companies with outgrowers need to cherish them and not give preference to production from their own fields. Both parties have to understand what the other party wants, if contracts are to succeed.

Warehouse receipt schemes are good in theory. But they are extremely complex to manage and require very high standards of efficiency and integrity. The jury is still out.

Case Study 9: Credit and Contracts – Contrasting Experiences with Cotton and Tobacco in Tanzania

Cotton

Cotton was grown in Tanzania in precolonial times, and then spun and woven to make garments. The Germans who colonised the country wanted large quantities of cotton. Their attempts to grow the crop on plantations were not successful, but they realised that small farmers could grow it. The British drew the same conclusion, but it was not till the Second Great War and the years after that it became a major export product, with most of the cotton grown in the area South and East of Lake Victoria. Its rapid expansion was assisted by the rapid growth of farmers' co-operatives, which purchased the cotton from the farmers, and sold most of it to co-operative unions who owned ginneries (where the cotton lint was separated from the seeds on which it grows). They in turn sold it to the Lint and Seed Marketing Board, which arranged for its sale and export overseas [12]. Cotton is a slow-growing crop, which attracts insects, some of which stain the crop or slow its growth. To control these, heavy doses of insecticides may be applied, and, where soils are lacking in nutrients, fertiliser. Credit, through which farmers could obtain these inputs, was arranged through the co-operatives.

The co-operative primary societies and unions in Tanzania were disbanded in 1976, and replaced by village committees and parastatal bodies ("crop authorities"). This created problems in the cotton-growing areas, because the village committees were not well suited to managing the distribution and repayment of credit. The villagers did not all grow cotton and many did not want to be involved in an arrangement in which they were collectively responsible for repaying loans. Co-operative primary societies and unions were reinstated in 1984, but as Joe Kabissa, a former Director General of the Tanzania Cotton Board, says they were not proper co-operatives: membership of primary societies was compulsory, and the unions had no assets and depended on loans from the Government and external donors and their senior managers were appointed by the Government [14, p.30]. The situation worsened in the 1980s when the structural adjustment policies that the country negotiated with the IMF meant that the prices farmers received for cotton did not keep up with the prices of goods in the shops. In the 1990s, ginning was opened to the private sector, and many small firms entered the market [13]. The result was overcapacity, and competition to purchase whatever cotton was available. This made it hard for ginneries to negotiate forward contracts with manufacturing companies, because they could not be sure of purchasing sufficient cotton. It led to a rapid decline in quality because ginneries were prepared to purchase any quality of cotton, even cotton that had been adulterated with sand or soil, or had water added, and there was no incentive for either farmers who grew the cotton or the traders who purchased it to reject poor quality cotton [10] [14]. It became almost impossible for ginners, or their agents, to offer credit, because it was easy for the farmers to sell their cotton to another ginner, and avoid paying back any loans they took on.

More recently, the Government, with the support of a private foundation, the Tanzania Gatsby Trust, tried to introduce contract farming, to improve the quality of the cotton. Small groups of farmers were encouraged to sign contracts which committed them to selling their cotton to a specific ginner. In return for this they got credit for the purchase of insecticides and fertilisers. This was piloted between 2008 and 2010 in the Mara area, East of Lake Victoria, where there were fewer ginneries, with some success. But when in 2011 the Government attempted to roll it out in the whole cotton growing area, there was a strong reaction from the owners of many of the smaller ginneries, who feared that they would lose their businesses if this continued, as they would not be able to compete with larger ginneries which were in a better position to negotiate contracts with overseas buyers. So the attempt to introduce contract farming, and make possible the provision of credit, failed [9]. The result is a very poor quality crop, and very poor prices. Cotton is no longer a major source of foreign exchange.

Tobacco

Flue-cured tobacco was not grown in Tanzania till 1940, and not by small farmers till the late 1950s, when it was introduced in the Tabora area. In 1958 a new design of barn made it possible for small growers to dry their own crop [10]. Production has grown steadily, and in recent years tobacco has been the country's most successful agricultural export.

Tobacco marketing is a complex process, because to get a high quality product the plants have to be grown with care to prevent damage to the leaves, which are then dried under precisely prescribed conditions, before being sold to a company and "cured" in a factory. The crop is then ready to be manufactured into cigarettes. Co-operative primary societies were involved in this from the start, arranging the credit for fertilisers and drying equipment, buying the leaves from the farmers, and selling them to the tobacco curing factory.

In this situation, the abolition of the co-operative unions and primary societies did not have such an adverse impact as it did with the cotton crop. Almost all the farmers in the tobacco areas grew the crop, and did not doubt that it would be difficult to grow it successfully without the purchased inputs. So the arrangements continued, more or less unchanged. When co-operative unions and primary societies were reinstituted, they were able to take back their functions as before.

Tobacco production is a form of contract farming. In Tanzania, there are three kinds of contract. The first, a single piece of paper, commits the farmer to selling the crop to the local primary co-operative society. In return the co-operative society agrees to purchase the crop and to provide credit for the inputs that are needed to grow it. The second contract is an agreement between the primary society and a bank. The bank agrees to make a loan to the primary society. That enables it to purchase the inputs which it supplies to the farmers, and to lend them cash to assist with the costs of the drying and grading. In return the members of the primary society agree, collectively, to repay the loan to the bank. The third contract involves the primary co-operative society, the co-operative union, and one of four tobacco processing companies that operate in Tanzania. The primary society, supported by the union which represents all the primary societies in the region, agrees to supply a given quantity of tobacco, and the company sets out a schedule of its grades for different colourations and thickness of leaves, and the prices it will pay for each. In 2012-2013 there were seventy-two such grades, each with a different price.

The two crops compared

The main difference is that in the case of cotton it has not been possible to control side-selling, whereas this is less of an issue with tobacco. There are many ginners willing to purchase the cotton, including what Joe Kabissa calls "marching guys", who come with a lorry in the middle of the night and purchase cotton without leaving any records of what they have done [14, pp.130–137]. They avoid paying taxes on the crop, and they give good prices for poor quality cotton, but they make it impossible for more reputable ginners to take the risk of providing inputs on credit, or to forecast how much cotton they will be able to buy.

In the case of tobacco, there are only four companies which process the crop, all externally owned. Two of these own processing factories, neither of which is in a tobacco-growing area; the other two use spare capacity in these factories or export the crop unprocessed. Each primary society contracts with one of the four. None of them have an interest in unofficial purchases. There are still some issues. For example some farmers purchase their inputs for cash, and do not take loans, and therefore get higher prices, and some larger farmers now have individual contracts with the processing companies. Some of the farmers who have taken credit may collaborate with these farmers to get their higher prices; but if this happens to a large extent it becomes obvious, and until now it has not destroyed the arrangements for the provision of credit. Farmers complain about the grading systems, which they feel are loaded against them, especially when they do not get the higher grades and prices. They feel that they are not given all the relevant information about what is needed to get the highest grades. They also complain that some of the inputs arrive late. These are real grievances, and it would be wise for those involved to be more transparent and to work more closely with the farmers. But they all accept that if they allow the contracts to be abused, and side-selling to become the norm, it will put the whole sector at risk.

This case study includes material also in the article by Kuzilwa et al [15], used with permission of the publisher.

Case Study 10: Dairy Quality Chain – Tanga Fresh Limited

The main sources for this case study are the 2006 Matchmaker Associates report on contract farming ([4], pp.26–8) and Marc Wegerif's PhD thesis ([11, also referenced in Chapter 7, pp.172–184]). Wegerif's work draws on the Masters dissertation of Rosanna Martucci [15].

Marc Wegerif writes:

> We sat on the porch of Mama Anna's house. A few metres away a calf was tethered and lying relaxed in the sun. Mama Anna has eight cows and three calves that graze on communal land and sleep at night in a small cattle *boma* (enclosure) next to her house. She milks the cows every morning and evening and carries the milk on foot to the Tanga Fresh collection centre. Her first cow, that she got more than 12 years ago, came from a 'take one give one' cattle project. She was given a cow and then paid back through passing on a calf to another family. She depends for her income on the milk sales and farming maize, oranges and coconuts on land in a neighbouring village where she was born. While she sells most of her milk to Tanga Fresh she also keeps some for home use and sells some to neighbours.
>
> Mama Anna has attended a two-week training on caring for dairy cattle run at a local college. When her cattle are sick she takes blood samples to a veterinary laboratory in Tanga for analysis and to get medicine. There is a vet in Pongwe who assists with artificial insemination. [Mama Anna] gets vitamins from the farmers' cooperative that runs the Tanga Fresh collection centre and buys *pumba* (maize bran) as supplementary feed from the maize millers in the same street in the village. When she has financial needs, often for family reasons rather than for dairy production, she borrows money from the milk collection centre and her loan repayments are deducted from the money she receives for her milk supplies. When needing advice Mama Anna does not feel she gets it from Tanga Fresh, but she explains how she does benefit from sharing experiences with other dairy *wakulima* [farmers] when they meet and talk at the collection centre.

The Tanga Dairy Co-operative Union (TDCU) was registered in March, 1993 as an apex organisation for nine dairy primary societies from five districts in Tanga Region (Tanga, Muheza, Pangani, Korogwe and Lushoto). In 2006, there were about 3,000 dairy farmers in this area, 40% of whom were women. Of the total, 1,500 were members of co-operatives. By 2015, there were 47 milk collection centres that took milk from around 6,000 small farmers.

The dairy company, Tanga Fresh, started in 1997, with an initial processing capacity of 15,000 litres a day at a factory in Tanga. In

2007 the Netherlands company DoB Equity invested €3 million in the enterprise. This made it possible for the factory to move to large modern premises, capable of processing 50,000 litres a day into fresh pasteurised milk, mtindi (sour milk), yoghurt and cheese; 80% of the milk is sold in Dar es Salaam, 340 kilometres by road from Tanga. By 2015, Tanga Fresh was employing around 150 staff. TDCU now holds 42.5% of the shares in this company, DoB owns 52.5% and the remaining 5% are held by two individuals associated with the company. (The role they played is described below.)

Farmers bring their milk by bicycle or cart to the collection centres each morning, with society members receiving preferential prices. The milk is chilled and then TDCU collects, transports and sells it to Tanga Fresh. TDCU has installed an accountancy system at the primary societies and monitors their accounts. They have also installed equipment to monitor the quality of the milk. The primary societies identify farmers whose milk is adulterated or of poor quality and follow up on the reasons for this, with assistance from Tanga Fresh and TDCU.

The Tanga Fresh concept is a unique success story in the dairy industry within and beyond Tanzania. It is the outcome of decades of interventions aimed at developing the dairy industry in the region. The Netherlands government supported large-scale state-owned dairy farming in Tanzania in the 1970s. That failed in the 1980s, but the Netherlands continued to play a leading role with more focus on the private sector and small-scale dairy farmers, through the Tanzania Smallholder Dairy Development Project (TSDDP). A wide range of initiatives was implemented, including research, training, improving cattle breeds, the "take a cow, give a cow" distribution of cattle, and the establishment of the Co-operative Union in 1993.

The TSDDP project came to an end, but the production and marketing system continued to function. Not least, this was because first the Co-operative Union and later TFL were reliable, paying for their milk on time. Initially, they paid on a daily basis because farmers did not trust the system. Now payments are deposited every two weeks into the primary societies' bank accounts.

The two key individuals who had the initial vision for the project saw it as a viable business that could also have a wider impact on development and poverty reduction. Lut Zijlstra led the Dutch development interventions in the 1980s and remains on the Board of Tanga Fresh. Alnoor Hussein, a Tanzanian national, was the first Managing Director and ran the company for more than a decade. He continues as a member of the Board and is very involved with the development of the dairy sector in Tanzania.

Some observations:

- Each primary society owns and operates a milk chiller, a machine costing about $10,000. These chillers were purchased under the Netherlands programme and given to the societies when the project ended. In 2007 they were replaced with larger and better-quality machines.

- Tanga Fresh sets a price that it pays farmers, and this stays the same for a whole year, even though there is much less milk in the dry season. It then sets recommended prices for its milk and other products, though most shops and outlets in Dar es Salaam sell for more than this recommended price.

- Milk buyers from other companies or individual traders offer higher prices during dry seasons than Tanga Fresh, and this encourages side-selling, which is against the rules of the co-operative. Even though they could, at times of shortage, sell their milk for higher prices, as a shareholder in TFL, TDCU is committed to its long-term relationship with Tanga Fresh.

- With the support of Tanga Fresh, TDCU provides inputs to its primary societies. These include vaccination, artificial insemination and financial services (soft loans) through Farm Friends Tanzania and Farm Friends Netherlands. Through this intervention, TDCU builds trust with its primary societies; however, the challenge for TDCU is to build the management capacity and leadership of these primary societies, so that they can respond to needs of the rapidly changing dairy market.

- Members of primary societies do not have separate contracts for selling their products to TFL. The Membership Agreement requires all members to sell all their milk to TDCU.

- The co-operative policy prohibits TDCU from intervening directly in primary society issues; for instance if the problem is leadership in a primary society, TDCU cannot intervene to rectify this. It is a matter for the primary society itself.

- The products – packets of pasteurised milk, yoghurt or cheese – are unfamiliar to most of the small farmers, and so are the systems required to maintain their quality. So although TFL makes big efforts to communicate with the farmers, in practice, given the layers of structures, the farmers feel removed from the leadership of Tanga Fresh and even from TDCU. One primary society leader complained that "they [TDCU] tell farmers they are the main shareholders, but farmers do not even know whether this year they ran at a loss or a profit" ([11], p.83). When there are profits they are dispersed according to the supply from each farmer during the year, not according to their ownership of shares in the company, so the farmers are motivated to produce more milk, while having little involvement in the affairs of the company.

Note: This report was accurate at the time of writing, but the situation can change. TFL has to compete with very large companies, selling milk which they "reconstitute" from milk powder which they import. And Tanga is a long distance from Dar es Salaam. TFL and its related co-operative union and primary societies have done very well so far, but the future is always uncertain.

Case Study 11: Cultivation of the Antimalarial Plant *Artemisia Annua* Under Contract to a Processing Company

This case study was written by Antony Ellman, co-author of the current book, who was one of the pioneer growers of Artemisia annua in Tanzania in 1994–95, and over the following twenty years undertook consultancies on cultivation and processing of the crop in East Africa, Madagascar, India, China and Vietnam.

Background to Artemisia cultivation in Africa

Artemisia annua is an annual plant, indigenous to China, where its medicinal properties have been known for more than two thousand years. Artemisinin, a chemical extracted from the leaves, was identified by Chinese scientists in the 1970s as being highly effective in treating malaria. Artemisia cultivation was introduced to East Africa only in the mid-1990s, when the demand for new antimalarial treatments had become an urgent priority.

High-yielding varieties of Artemisia annua were developed in the 1980s and early 1990s by Mediplant, a medicinal plants research institute in Switzerland. Seeds from Mediplant were obtained by a group of entrepreneurial farmers and engineers in Tanzania and Kenya, and in 1994 trial plantings were established at a number of locations in both countries. The crop grew well, so the entrepreneurs established a company in Tanzania, called African Artemisia Ltd. (AAL)12, to promote its cultivation and to establish factories, initially in Kenya and Uganda, to process and market the product.

Contract growing of the crop

Since AAL had no land of its own on which it could grow Artemisia, its best option was to encourage local farmers to grow the crop on its behalf. The company would then buy the dried leaves for processing in its factories. Artemisia is not a difficult crop to grow, but some stages of its production – notably transplanting, weeding and harvesting – require a lot of labour and attention to detail, making it an ideal crop for smallholder producers.

As Artemisia had not previously been grown in East Africa, it was necessary for AAL to provide farmers with seed or seedlings on credit, to give advice and training on the best cultivation methods, and to offer

---

[12] AAL changed its name to Advanced Bio-Extracts when it expanded its operations from Tanzania into neighbouring countries.

a guaranteed market for the harvested leaves at a price which made Artemisia cultivation competitive with other crops that the farmers could grow.

Farmers needed to be persuaded to try out this new crop. So the services which AAL would provide, and the actions which would be expected of the farmers in return, had to be spelled out in a clear contract defining the rights and obligations of company and farmers, as well as the penalties which could be imposed if either side failed to fulfil its obligations, and the bonuses which would be offered as an incentive if the farmers did more than was expected of them.

The Artemisia contract

A contract was drawn up between AAL and its contracted Artemisia growers. At the time it was a model of its kind, with the following components:

a) Obligations of the company:

   i) AAL agreed to supply the farmer with seed and other necessary inputs on credit, recovering the cost from the value of the leaf delivered by the farmer;

   ii) AAL also agreed to advise the farmers on Artemisia cultivation methods without extra charge;

   iii) AAL guaranteed to buy all the farmer's Artemisia leaf of agreed quality, at a price calculated according to the formula outlined below.

b) Obligations of the farmer:

   i) The contracted farmer agreed to grow Artemisia on land confirmed as suitable by AAL, and to follow the technical advice given by AAL staff;

   ii) The farmer agreed to sell all harvested leaf of the required quality to AAL, and not to sell leaf to any other buyer (since there were no other buyers of Artemisia leaf at the time, this was not a significant issue).

c) Formula for calculating price paid for Artemisia leaf:

   i) AAL initially purchased all leaf, sifted to remove stems and other materials and dried to a moisture content of less than 13%, at a price of US$250 per tonne.

   ii) A second payment of US$100 per tonne of leaf was offered to the farmers after extraction was completed, provided the artemisinin content of the leaf was above a minimum of 0.5% by weight;

iii)   A further bonus payment of US$40 per tonne was paid for each additional 0.1% of artemisinin content above the minimum of 0.5%.

The base price of US$250 per tonne of leaf was calculated at a level which gave the farmers an acceptable return per hectare compared to other cash crops such as maize, green beans, sunflower, potatoes and coffee. It was also calculated to allow AAL to make a profit on extraction and sale of artemisinin at the price prevailing at the time.

If farmers delivered leaf below the required quality, they forfeited the second payment and the bonus payments. If AAL failed to meet its obligations to the farmers, compensation would be offered but this has not had to be put into effect.

Results of the experiment

There was initially great enthusiasm among small farmers for growing Artemisia in East Africa, especially in Kenya and Uganda where processing factories were located, but also in Tanzania where AAL started its operations. Some farmers even went as far as to uproot old coffee bushes to make room for Artemisia, since the coffee price at the time was very low.

At its peak in around 2009, some 2,500 hectares of Artemisia were planted in Kenya, Tanzania and Uganda by around 5000 small-scale growers, and there was great satisfaction with the outcome.

Unfortunately, from 2011 onwards the experiment encountered increasing difficulties, largely because of a rapid expansion of Artemisia cultivation in China, taking advantage of the high (if fluctuating) price for artemisinin. The price for artemisinin produced in East Africa therefore fell below an economically viable level. First the factory in Uganda (a converted pyrethrum factory) closed down, then the Kenya factory failed to pay its contracted farmers in Tanzania and Uganda, so farmer confidence in the enterprise was lost, and the enterprise itself collapsed.

Thus the East African experiment, though it started out well, was ultimately not a success. However, a similar enterprise in Madagascar, started by an agro-industrial company which processes a range of agricultural crops as well as Artemisia, still contracts over 10,000 small-scale Artemisia growers and runs a flourishing contract farming industry. The main reasons for its success are that the company is not totally dependent on artemisinin sales, and also that it has negotiated favourable forward sales for artemisinin. Both these factors cushion it against periodic down turns in the artemisinin price.

Analysis of the contract farming experience

While the attempt to introduce Artemisia cultivation to East Africa has not been a long-term success, this experience illustrates the value that can be derived from a well-designed contract farming relationship between growers and buyers of a profitable agricultural commodity.

The transparency of the price formula designed by AAL, which aimed to share the profits of the enterprise equitably between growers and processor, played a big part in its appeal to farmers. The bonus payments for high quality leaf, linked directly to the profitability of the overall enterprise, gave farmers a stake in the company's success and a strong financial incentive to contribute to it.

Provided that a sure market and an attractive price for the end product are confirmed, companies aiming to contract farmers to grow other new crops in Africa can benefit from the Artemisia experience described in this case study.

Further reading on open access on the internet

1. Nicholas Minot (2007) 'Contract Farming in sub-Saharan Africa: Patterns, Impact and Policy Orientation'. Case Study 6.3 in *Case Studies in Food Policy for Developing Countries: The Role of Government in the Global Food System* Volume 2, edited by Per Pinstrup-Andersen and Fuzhi Cheng, Cornell University 2009. https://cip.cornell.edu/DPubS/Repository/1.0/Disseminate?view=body&id=pdf_1&handle=dns.gfs/1200428173

2. DEFRA (2017) *Food Statistics Pocketbook 2016* London: Department of the Environment and Rural Affairs. https://www.gov.uk/government/uploads/system/uploads/attachment_data/file/553390/foodpocketbook-2016report-rev-15sep16.pdf

3. Charles Eaton and Andrew Shepherd (2001) *Contract Farming: Partnerships for Growth*. FAO Agricultural Services Bulletin 145. http://www.fao.org/docrep/014/y0937e/y0937e00.pdf

4. Matchmaker Associates (2006) *Contract Farming: Status and Prospects for Tanzania*. Final Report for the Ministry of Agriculture Food and Co-operatives Participatory Agricultural Development and Empowerment Project. http://www.fao.org/uploads/media/Contract%20farming_Tanzania.pdf

5. Yuko Nakano, Yuki Tanaka and Keijiro Otsuka *Can "Contract Farming" Increase Productivity of Small-Scale Cultivation in A Rain-fed Area in Tanzania?* GRIPS Discussion Paper 14–21 2014. http://www.grips.ac.jp/r-center/wp-content/uploads/14-21.pdf

6. The Warehouse Receipts Regulatory Board (of Tanzania). http://www.wrs.go.tz/

7. J.P. Coulter, and Andrew Shepherd. *Inventory Credit: An Approach to Developing Agricultural Markets*. FAO Agricultural Services Bulletin No. 120, 1995. ftp://ftp3.us.freebsd.org/pub/misc/cd3wd/1005/_ag_inventory_credit_agric_marketing_unfao_en_lp_108940_.pdf

8. Common Fund for Commodities (CFC). 2006. *Mfumo wa stakabadhi za Maghala. Mradi wa Kuendeleza Masoko ya Pamba na Kahawa*, Dar es Salaam, Tanzania. www.wrs.go.tz/downloads/newsletter/mfumo.pdf

9. Poulton, Colin. *APRA Policy Processes and Political Economy: Tanzania Country Review*. Working Paper 5, Future Agricultures Consortium, pp.26–9, 2017 https://opendocs.ids.ac.uk/opendocs/bitstream/handle/123456789/13561/APRA%20W.P.%20number%205%20web.pdf?sequence=1&isAllowed=y

10. Boesen, Jannik and A T Mohele *The "Success Story" of Peasant Tobacco Production in Tanzania: The Political Economy of a Commodity Producing Peasantry*. Uppsala: Scandinavian Institute of African Studies, 1979. https://trove.nla.gov.au/version/30974735

11. Marc Wegerif *Feeding Dar es Salaam: a symbiotic food system perspective* PhD thesis, Wageningen, 2017. http://edepot.wur.nl/414390

References not on open access on the internet

12. Coulson, Andrew *Tanzania: A Political Economy*. Second Edition. Oxford University Press, 2013 (first edition 1982).
13. Pete Gibbon "Free competition without sustainable development? Tanzania cotton sector liberalization 1994/95 to 1997/98". *Journal of Development Studies* No.36 Vol.1 pp.128–150, 1995.
14. Joe Kabissa, Joe (2014) *Cotton in Tanzania: Breaking the Jinx*. Tanzania Educational Publishers and Africa Book Collective.
15. Joseph Kuzilwa, Bahati Ilembo, Daniel Mpeta and Andrew Coulson. *Contract Farming: Experiences from Tobacco and Sunflower in Tanzania*. In David Potts (ed.) *Tanzanian Development: A Comparative Perspective*. James Currey. 2018
16. Martucci R. (2015) *Integrating small farmers in liberal economy*. The case of Tanga Fresh in Tanzania. Masters Dissertation, International Development Studies, University of Amsterdam.

Topics for essays or exam questions

1. What do you understand by the term "side-selling"? What actions can banks and governments take to make it difficult or to prevent it?
2. In developed countries, most lending to farmers is secured against the farm (so the bank can take possession of the farm if the farmer does not repay as agreed). Discuss the possibilities of increasing lending to small farmers in Africa secured in this way.
3. Imagine you are a banker, and you are approached by someone who wants to start a business that will purchase crops from small farmers. What topics would you expect to be discussed in the business plan, and what questions would you ask before approving a loan for short-term finance?
4. Discuss the possible uses of M-pesa to provide loans for small farmers.
5. Discuss the issues that arise in providing credit for the purchase of (a) moving assets and (b) fixed assets.
6. Explain the term "contract farming". What are its advantages and disadvantages? Explain why contract farming has often failed in Africa.
7. Contract farming is more likely to succeed if the crop concerned is perishable, because side-selling is more difficult with a crop that deteriorates over time. Explain this in more detail, and also explain why such contracts have often not survived in African countries.

8. Discuss why the products grown by small farmers are often of poor quality, and how contract farming may lead to higher quality of produce. What are the difficulties in making this happen?

9. Explain why contracts between farmers and processors of agricultural products are often unequal. What can be done to give the farmers more power and influence when they negotiate contracts, or if there are disputes afterwards?

10. Discuss why, over time, contracts often go wrong and trust is lost. Is the best remedy for this more involvement of farmers in the preparation and monitoring of the contracts?

11. Comment on the Tanga Fresh project described in Case Study 1. What are its strengths and its weaknesses?

12. How does a warehouse receipt scheme work? What are the possible advantages and disadvantages of warehouse receipt schemes in Africa?

13. Go to "Downloads" on the website http://www.wrs.go.tz. Look at the list of forms, and the WRS Operational Manual. Discuss whether this document is a realistic means of making this system work as planned in remote villages

CHAPTER 9

Agricultural Extension: Getting New Technologies to Where They are Needed

Key themes or concepts discussed in this chapter

Chapter 4, the last chapter in Part 1 of this book, was about research – the creation of new innovations, or the use of existing technologies in new situations. This chapter is about the spreading of new technologies – dissemination. Here are the key themes:

- It is not necessary to have extension workers for innovations to spread. Examples where innovations have spread without extension workers are the cultivation of round potatoes in Njombe, and the use of plastic hosepipes for irrigation in upland areas of Tanzania.
- Some innovations spread slowly. Others are so obviously beneficial that they spread very quickly.
- Most countries employ specialists to advise farmers on agricultural practices. But are the terms "extension worker" or "bwana shamba" good ways to describe them?
- An extension service may be organised in at least seven ways, through:
  o village-based generalists
  o crop specialists
  o sales staff in the private sector
  o specific projects, often funded by donors or NGOs
  o co-operatives or groups of farmers
  o research institutes
  o a form of policing, directing farmers what to do or not to do.
  There are problems with all of these.
- A deeper critique is that extension is essentially top-down. Often it does not recognise that the farmers already have a great deal of knowledge, and that the best way to assist them is to build on that knowledge. This would require institutional arrangements very different from those in most conventional research stations or universities.

- Many farmers innovate – they try out new techniques, or machines, new crops, or new methods of growing existing crops. The future lies with them. If they can be helped to innovate, and their innovations spread widely, then much greater production is possible from small farms.
- Extension is expensive, and often ineffective; so should the service be disbanded? That may be a step too far, but it is time for hard questions to be asked and for much more attention to be given to extending the knowledge already held by farmers.

MARKET FORCES – A POWERFUL DRIVER OF INNOVATION

The dissemination of an innovation may not need extension workers at all. At the end of this chapter there is a short case study of the spread of potato cultivation in the Njombe area. Historically, the land near Njombe was so unproductive that many farmers could not earn a living from agriculture, and migrated long distances to work on sisal estates or large farms. Some of these farmers came home with improved varieties of potatoes and planted them. Others got hold of new varieties when Uyole Agricultural Research Institute carried out trials on their farms. Either way, the Njombe farmers liked the results, and the crop was easy to grow – indeed it grew on its own because it was very difficult to harvest all the potatoes and some were left in the ground and produced a crop the following year. Farmers copied from each other, and spread the new varieties. This was not planned, nor a major national policy. Nor were potatoes the only crop that could be grown. Hybrid maize, supported by the use of fertilisers, tomatoes, onions, black wattle and pine trees, were also found to be successful. Njombe became one of the most productive agricultural areas in Tanzania.

Another example of an innovation that spread through market forces was the use of plastic hosepipes for irrigation in hilly areas. Traditionally furrows were dug to take water from one place to another, some with aqueducts constructed in stone to take the furrows across valleys. But today there is a kind of spaghetti of criss-crossing plastic pipes leading to sprinklers to distribute the water.

Geographers study the diffusion of innovations. Some innovations spread slowly, for example where the innovation depends on one farmer copying a neighbour. Other innovations spread like a fire, where as many people as can afford it adopt the innovation, wherever they are – an example is the spread of mobile phones in Africa. Rural residents did not wait for masts to be built: if necessary they climbed trees or went up the nearest hill to get a signal. There was no need for extension workers to spread the word about mobile phones!

A process of agricultural innovation may be assisted by radio or TV programmes that describe a successful new activity, and encourage farmers to try it. Such programmes may be sponsored by people with a commercial interest in a technology, such as a company which sells machines or crop sprays. Or by NGOs which promote new ideas. An example of this was the Rural Livelihood Development Company, an NGO supported by the Swiss government, which promoted the introduction of hybrid sunflower seeds in the Dodoma and Kongwa areas – backed up by contracts that would enable farmers to get access to the seeds, fertilisers and sprays needed for the best results. Social media may have a contribution to make: Facebook and YouTube can demonstrate what needs to be done, WhatsApp can enable groups of farmers to share their experiences and confront problems together – and perhaps also to learn where the best markets are. In many villages there are shops that sell products such as improved seeds, fertilisers, sprays, machines and drugs. Their proprietors have an interest in selling them, and hence in promoting the technology involved.

EXTENSION OFFICERS, OR ADVISORS?

In contrast to the examples described above where innovations were spread by market forces without assistance from the government or other agencies, the conventional approach to disseminating new agricultural technologies in Africa is to employ "extension workers". These workers may have different levels of expertise: in developed countries where farming is big business and many farmers have degrees in agriculture, extension workers are often graduates and many are the children of farmers or have been farmers themselves. At the other extreme, in Africa in colonial times extension workers were often little more than policemen who told farmers what to do and imposed fines or prison sentences if they did not comply. At the present time in Africa most extension workers have a certificate gained from two years of study after Form IV. This is better than just a few months of training, but it still means that they have only limited knowledge of the science lying behind a recommendation they make, and of whether it is appropriate in all situations. There will almost certainly be farmers who, because of their experience, know more about the likely consequences of an innovation or of how to respond to a pest, than the extension workers do.

The word *extension* derives from a practice in the USA where "extension services" were set up in universities to "extend" their work into deprived communities. When the Federal Government created a network of research institutes, it linked them to universities where agriculture was studied, and extension workers were employed to take recommendations and innovations to farmers in each area. In the UK, they are called "agricultural advisors". But other countries have better words to describe what extension workers do:

- Arabic: *Al-Ershad* ("guidance")
- Dutch: *Voorlichting* ("lighting the way ahead")
- German: *Beratung* ("advisory work")
- French: *Vulgarisation* ("popularisation")
- Spanish: *Capacitación* ("training" or "capacity building") or *Cooperante* ("partner", or "worker from outside")
- Persian: *Tarvij & Gostaresh* ("to promote and to extend")[13]

At Sokoine University of Agriculture in Tanzania, the Department of Agricultural Extension is *Idara ya Ushauri wa Kilimo* (literally "Advisory Department in Agriculture"). This is better than *Bwana Shamba* ("Mr Agriculture") which implies that those concerned are male: it also has authoritarian overtones.

The US practice is a working demonstration of what is often presented as a triangle: research – education – extension. The government supports the extension service by giving special grants to the universities involved. The professors teaching agriculture then work closely with the extension workers. The extension workers tell the researchers about the problems facing the farmers and give them new challenges for research. This is good in theory. But it has been difficult to make it work in Africa, probably because universities are autonomous and their researchers give priority to their students and to papers that can be published, not necessarily to work that can be directly communicated to farmers.

THE ORGANISATION OF AN EXTENSION SERVICE

The key to understanding extension is to realise that it may be organised in at least seven different ways:

1. In most African countries the largest employer of extension agents is the government – or, in Tanzania's case, local government. Most

---

[13] This paragraph draws on the Wikepedia entry for Agricultural Extension, accessed 6 June 2018.

of these extension agents are *generalists*, living in a village and expected to give good advice in relation to all crops or situations in the area.

2. They may be employed to *promote specific crops*, such as cotton, tea or tobacco. When companies use outgrowers, they almost always employ extension workers to assist the farmers, and sometimes also to liaise with the parent company.

3. Extension agents may also be employed in the private sector, as *salesmen or women*. They are expected to arrange the sale of fertilisers, insecticides, seeds, machinery or other purchased inputs. Others run small shops which sell seeds and agricultural inputs.

4. They may be employed by a *specific project*, supported by an external donor or an NGO – often to bring about a specific innovation (for example, contract farming among sunflower growers in Central Tanzania) – or by the World Bank, which funded large projects in the 1980s promoting crops such as maize.

5. They may be employed and paid for by a *group of farmers or a co-operative* who wish to develop a new crop – though for small farmers this may be too expensive without support from the state.

6. They may be employed by, or through, *research institutes or universities*. This is how extension work on coffee and tea is presently funded in Tanzania.

7. *Extension as policing*. Extension agents can have legal, or semi-legal, powers. For example they can fine, or report, or in some way punish farmers who do not burn the remains of their cotton crop at the end of a season, or do not remove grass growing under cashew trees.

None of these methods of organising the service is without controversy.

Generalist extension workers are expensive to employ, and difficult to supervise effectively, so it is often not clear exactly what they do day-to-day, or what contribution they make. It is very hard to find a performance measure to evaluate the contribution of a generalist extension worker.

One response to this is to *concentrate the extension workers on a few farms*. Thus some of the World Bank projects proposed to employ one extension worker for a hundred farm families. On that basis they could visit each farmer every few weeks. But this is an expensive overhead: is it realistic to envisage that these extension workers could increase the production of one hundred farmers sufficiently to pay for their own salaries, not just for a short time but year after year? The principle was taken to its logical conclusion in the "Training and Visit" system popularised by the World Bank in the 1970s and 1980s. Here the extension workers were brought together for training at regular intervals, and agreed on key extension messages to be communicated to farmers. Farmers agreed to meet the extension workers every two weeks, and to carry out their recommendations. The system was top-down: the extension workers came with recommendations and the farmers were expected to implement them. In some countries, the extension workers were all men, even though much of the work in the fields was undertaken by women. The system broke down partly because of the difficulty of formulating recommendations which would be generally relevant, and because farmers stopped attending the meetings, but above all because of its expense.

If *extension workers work on a single crop*, they may have no interest in other crops and no understanding of the interactions between crops (discussed in Chapter 5). If they work purely on crops they may miss the important contributions of livestock – and vice versa. This is reasonable where a crop such as tea, tobacco, sugar or sisal dominates a farm, as with some outgrower systems, but not for the typical family farm with many crops and livestock.

Salespeople in the private sector are often paid a commission– usually a proportion of the money they bring in to their employer. This is likely to increase the quantities of products sold to the farmers in the area, even if they are not the cheapest, or the most suitable. Other salespeople may be employed by a company that purchases a particular crop, such as tobacco. The use of inputs may be built into contracts – which is fine provided the recommendations are appropriate to the particular crop and area.

Project-based extension usually ceases when a project ends, and *co-operative groups* often find the costs of employing salaried workers prohibitive. The problems of relating extension to *research institutes or universities* have already been discussed, though some individual academic staff are very enthused by working with farmers or groups of farmers, and this can be valuable for those involved, even if the numbers are small.

There are regulations which have to be enforced so that ultimately the whole community benefits. For example the prohibition on cutting down trees near springs. If the trees are cut down soil erosion is very likely, and the springs may easily dry up. There is a requirement that cotton bushes be burnt at the end of a season, to prevent any diseases on the plants reappearing the following year. Cattle should be dipped regularly to control ticks and tick-borne diseases. There may be regulations about the use of water, designed to ensure that limited supplies of water from streams are shared fairly. In colonial times, much of the extension advice was based on compulsion: farmers were forced to grow minimum areas of certain crops, to protect steep hills from erosion, to limit the numbers of cattle in an area, and so on. Extension workers enforced these regulations – a form of policing. Some enforcement is necessary, but if extension workers are seen as police they are not likely to gain the trust of the farmers, nor will they be able to help much with the more fundamental problems of increasing agricultural production. To win their main arguments by discussion, not by force, extension workers have to be confident, and sufficiently well trained.

In the last resort, extension depends on trust – on workers being able to explain new technologies in simple terms, and farmers being willing to listen, to consult them, and to experiment.

A DEEPER CRITIQUE OF AGRICULTURAL EXTENSION

The discussion above is a very strong critique of the different ways of organising extension work. But there is a deeper, more profound critique. The concept of 'extension workers' is, at its heart, top-down. "In a Transfer of Technology approach, the extension agent is in the powerful controlling, position, being the purveyor of 'new' knowledge to farmers" ([2], p.6). The messages and innovations come from the extension workers or the technologists who train them; the farmers are passive recipients who are expected to do what they are told.

This was the fundamental criticism of the educational philosopher Paulo Freire who worked with farmers in rural Brazil. He criticised the mind-set that lies behind the assumption that the extension workers know what is good practice, and so their job is to tell the farmers what to do.[3] This does not recognise that the farmers already know a great deal about how to grow crops, including what *not* to do. In fact the farmers may have greater knowledge than the extension workers, and many of them have heard the same advice many times before – for example that they should use more inorganic fertilisers – and have decided not to adopt it. If this is the case, the farmers probably have logical reasons for their behaviour. Freire advocated *dialogue* – listening to farmers, and responding to their problems, and if necessary changing the recommendations. He preferred the term *co-operante* (advisor, or partner, or co-operator) – one who works with the people, not just one who tells them what to do. He talked about *conscientization*, in which the people in a community – in this case the farmers – become conscious of how they are oppressed and exploited, but also see how to obtain better techniques, through technology, and become better able to influence markets, and to improve their position.

Freire's critique was turned into practical policies by Robert Chambers in his concept of "farmer first" (introduced in Chapter 4 above) as an approach to research [8]. It is also a philosophy of extension. It means that what happens to assist farmers should start with the farmers themselves and their problems – not with extension workers or research specialists. If farmers can identify problems, then extension workers, or researchers who support them, should be able to suggest possible solutions or improvements. Farmers should be encouraged to try out these suggestions, learn from them, and then teach other farmers about what is involved.

This approach also recognises that advice is "contingent" – it should change all the time in response to circumstances. Thus if the price of one crop goes up and another goes down, the advice should change. If the rains are late, the advice should change. If there is a shortage of a particular fertiliser, the advice should change. The extension worker should not be conveying a fixed set of messages or techniques – as, in the extreme case, with the "Training and Visit" type of extension – he or she should be altering the messages all the time as circumstances change.

In a subsequent book, *Beyond Farmer First* [9], Ian Scoones and John Thompson take the argument even further. In the Introduction [2] they argue that simple collaboration still leaves the experts in a position of power, because it is they who propose the solutions. But close working with farmers and groups of farmers will show that they have huge amounts of knowledge – about ways of minimising or responding to risks, different shrubs and products of the forest, the maintenance of soil fertility, the ways to use different soils, and how to select the best seeds for the following year. So solutions should come at least as much from the farmers themselves imparting their knowledge, as from professional outsiders.

Scoones and Thompson argue that most agricultural research, especially that carried out at agricultural research institutes, is far more simplistic than the real world ever is. Thus a farmer may have a broad plan about what will happen on the farm during the course of a season, but what actually happens will depend on when the rains come, how strong they are, changes in prices, and possible outbreaks of pest infestations and diseases – to take just some possible variables. Conditions on a farm will not be the same as they were the previous year, and the farmers will be constantly adapting what they do in the light of circumstances developing around them, whereas research usually concludes with a single "one-size-fits-all" recommendation – usually for the actions that will produce the highest yields.

The reality is much more complex. Some farmers are more knowledgeable than others. Some have had more opportunities to experiment than others. Some people like to try out new ideas. Some are so poor that they can barely survive, and cannot afford anything that might make things worse. There may be divisions within a family, in which some decisions are taken by men, others by women, and that too may make innovation harder. Increasingly farmers are literate and can use mobile phones and the internet, but some can only use verbal – and therefore local – forms of communication. It is not a coincidence that much extension work takes place with bigger, better off, farmers. They have confidence, and can afford to experiment, and are easier to talk to. Nevertheless some of those with the greatest knowledge of the secrets of the land will not be literate or have large farms.

ANOTHER APPROACH: IDENTIFYING INNOVATORS AND THEIR INNOVATIONS

It is not true that all small farmers do not innovate. Reij and Waters [1] identified "about 1000 farmer innovators and concluded that innovation is a fairly common phenomenon in regions where there is high population pressure on available natural resources". Some of the innovations involved buying expensive machinery, and could only be implemented by large farmers. Others involved very simple changes – to the way things were done, the order in which they were done, the materials used – and could be adopted by anyone.

So a completely different approach would be to start by identifying the innovating farmers in an area, and then taking steps to diffuse their innovations more widely, through meetings, visits, mobile phone technologies (such as WhatsApp), or through training programmes for other farmers.

Specialist research workers would be in the background, available as consultants, or ready to make suggestions. They would also continue to carry out long-term projects of plant and animal breeding, but any resulting improvements in the seeds and planting material would be those selected and desired by farmers, not necessarily by "experts" or international organisations. This approach would move extension work away from the dissemination of a few messages centrally created, to a flexible approach where the messages are generated in the community, improved in research institutes, and then disseminated largely by farmer-to-farmer contact.

CONCLUSIONS

Should Tanzania Disband its Extension Service?

The discussion above has shown that there are two fundamental problems with the extension service:

1. Without a close working relationship with farmers on the one hand and an active research service on the other hand, a generalist extension service is likely to become an expensive luxury. Extension workers will probably not have up-to-date messages or information. Even though they may play a useful role in a village – taking the minutes at meetings, welcoming visitors, lobbying at the District Office on behalf of villagers, and so on – they are unlikely to make much contribution to agriculture.

2. Even if there are good relationships between farmers, extension workers and researchers, these may be one-sided and ineffective because all the power and initiative lies with the researchers and the extension officers, and very little power with the farmers. There may still be some good outcomes – for example, varieties of seeds or planting material that are resistant to diseases may be released. But the messages conveyed may not take account of the knowledge of the environment within the community, and farmers may have fundamental – and justifiable – reasons for rejecting the suggestions. Extension and research are likely to work much better if farmers control the agenda, or at least have a big influence over the research (see [4] and the interesting papers in [5]).

As noted at the start of the chapter, it is not necessary to have an extension service for innovations to spread. A good innovation will spread on its own, by word of mouth, farmer-to-farmer contact, and through market forces.

To disband the present generalist extension service in a country such as Tanzania would be a big step. But it might help to have somewhat fewer village-based generalist extension workers, and to use the resources so liberated to pay for better transport, more farmer training and dialogue, and regular visits to research institutes.

Above all, means must be found through which the extension service becomes the servant of the farmers in an area – helping to answer their questions and expand their opportunities – and not a service which is set up to tell farmers what to do. Ultimately it is the knowledge already held by farmers which is the key to successful innovation in the future, and the purpose of extension should be to help this happen – and then to spread the good news.

Case Study 12: Cashew Nuts in Vietnam and Tanzania

In the early 1980s, the government of Vietnam decided to develop a cashew nut industry. The story of how they did it is told by Blandina Kilama in her PhD dissertation [6].

The cashew plant was introduced into Vietnam in the socialist period, mainly as a means to limit soil erosion. But after the Vietnam War, the Vietnamese realised that they needed new sources of foreign exchange. They expanded production from almost nothing in 1988 to more than twelve times Tanzania's production by 2008 ([6], 127–8). They first created the capacity to process imported cashew nuts. This is not easy because the liquid in the shells of the nuts is highly toxic to hands and other parts of the body, and the nuts are easily broken. So they created a new processing machine in which the nuts are held firm in holders controlled by the operator's feet – a big improvement on holding the nuts in the hands. Once they had established efficient processing, there was a local outlet for raw cashew nuts. Only then did they start recruiting farmers to grow the trees. They were given good quality cuttings to plant, clear instructions about how far apart to plant them, how to prune them, what fertilisers and insecticides to use, how to harvest them, etc.

The result has been an intensive agriculture – with at least twice the yields achieved in Tanzania. Many of the Vietnamese trees are small "dwarf" varieties, quick to grow, easy to harvest, and highly productive. They are pruned to keep them small and sprayed to combat plant diseases. Later the Vietnamese developed improved automatic processing machinery, which was much better than the Italian technology used in some of the early processing factories in Tanzania, and much cheaper to purchase.

In Tanzania, processing was inefficient and marketing badly organised, both before and after liberalisation. Kilama's careful analysis (pp.45–56) shows how the inefficiencies of both processing and marketing meant that farmers got very poor prices. In the 1970s, Tanzanian farmers were forced to live in villages; many farmers had to move a long way from their trees, and so stopped looking after them. That led to the growth of weeds under the trees, and the spread of plant diseases which proved difficult to control.

The result is a very efficient sector in Vietnam, a much less efficient cashew sector in Tanzania. Da Corta and Magongo [2, Ch.6] showed that it was in fact possible to get yields in Tanzania similar to those achieved in Vietnam, but the farmer they studied chose not to continue this high productivity cultivation.

Case Study 13: Potatoes in Njombe: Dissemination Without Extension

The cultivation of round (or Irish) potatoes in the Southern Highlands of Tanzania has been a success. But most of that success is a result of farmers' initiatives and opportunities, with some involvement of researchers at Uyole Research institute near Mbeya.

An anthropological study published in 1996 [10] tells the stories of three farmers who were migrant labourers in the Arusha and West Kilimanjaro areas where they worked on farms that grew potatoes. They brought back some of these potatoes to the Southern Highlands and planted them, in one case as early as 1961, and they grew well. Fifteen varieties of potatoes being grown in the area in the 1990s were imported in this way. These were varieties originally bred in Kenya with resistance to blight. Several of these varieties are still being grown, including the red-skinned Irika variety in the highest areas, and the high-yielding Kagiri variety, which is suitable for boiling or mashing, in lower areas. When new varieties are introduced in Kenya, they quickly find their way, without state or company involvement, across the border to Tanzania and on to the Southern Highlands ([10], pp.91–92).

In 1975 a research programme was started at the newly opened Uyole Agricultural Research Institute. Varieties were created that were high yielding and resistant to diseases. Trials were set up on farmers' farms, and six varieties were selected. But in a follow up study in 1996 the researchers could not find any farmers who were still growing these varieties (p. 87).

The 1996 study related the expansion in the growing of round potatoes to a number of factors. From the early 1960s, pyrethrum had been a profitable cash crop in the highland areas. But its market collapsed in the 1970s, around the time that the road from Mbeya to Dar es Salaam was tarred. There were many trucks from Tanzania returning from Malawi or Zambia with spare capacity, making it cheap to transport potatoes to Dar es Salaam. The expansion started. Then in the 1980s, the production of cassava fell nationally, due to infestation by mealybugs, and in Dar es Salaam chips made from round potatoes partly replaced roasted cassava.

In 2010 another study found that 58% of a small sample of farmers growing potatoes in a ward near Mbeya were using an improved variety, 80% were using fungicides, 52% were using insecticides, and 90% were following the recommended density of planting. All the farmers were planting at the recommended time. Only 30% (18 farmers) appeared to be using fertilisers. So the work of the research institute and the extension service appeared to be having an impact in this area, though perhaps not as great as the researchers would have liked [11].

The production of potatoes in Tanzania has grown rapidly. The quality is good, and as a result potatoes have become an established part of the diet in cities such as Dar es Salaam. They are easy to grow, on ridges, on light soil on gentle slopes. A highly successful enterprise has been created and shows no sign of lessening in importance.

This is a green revolution. Research has done its work – much of it in Kenya – and farmers have learnt how to grow the crop, either from extension workers or from other farmers. They have adopted some recommendations (for example the use of fungicides) more enthusiastically than others (for example the use of fertilisers). Many of the preferred varieties of potato have been introduced by the farmers themselves on the basis of a range of factors, largely without assistance from the extension service. Market forces, and farmer-to-farmer contact, did most of the dissemination.

Further reading on open access on the internet

1. Chris Reij and Ann Waters-Bayer (eds.) *Farmer Innovation in Africa: A Source of Inspiration for Agriculture*. Earthscan, 2001. A short summary of this book is at https://cgspace.cgiar.org/handle/10568/63785
2. Ian Scoones and John Thompson "Knowledge, power and agriculture – towards a theoretical understanding". Introduction to *Beyond Farmer First* (see [9] below) https://www.researchgate.net/publication/300257835_1_Introduction_Knowledge_power_and_agriculture_-_towards_a_theoretical_understanding
3. Paulo Freire, *Pedagogy of the Oppressed*. Continuum 2015 http://commons.princeton.edu/inclusivepedagogy/wp-content/uploads/sites/17/2016/07/freire_pedagogy_of_the_oppresed_ch2-3.pdf
4. O.T. Kibwana, *Forging Partnerships with Innovative Farmers in Tanzania*. Links Project, FAO, 2003 http://www.fao.org/fileadmin/templates/esw/esw_new/documents/Links/Publications_Tanzania/31_Rep_3_Kibwana.pdf
5. Triomphe B, Waters-Bayer A, Klerkx L, Schut M, Cullen B, Kamau G & Le Borgne E (eds). 2014. *Innovation in smallholder farming in Africa: recent advances and recommendations. Proceedings of the International Workshop on Agricultural Innovation Systems in Africa (AISA)*, Nairobi. CIRAD 2013 http://aisa2013.wikispaces.com/file/view/AISA%20workshop%20proceedings%20final%20March%202014.pdf/497465208/AISA%20workshop%20proceedings%20final%20March%202014.pdf
6. Blandina Kilama, *The diverging South: comparing the cashew sectors of Tanzania and Vietnam* PhD dissertation, Leiden University, 2013, especially pp.27–63 and p,146 https://openaccess.leidenuniv.nl/bitstream/handle/1887/20600/fulltext.pdf?sequence=12
7. Marc Schut et al. "RAAIS: Rapid Appraisal of Agricultural Innovation Systems (Part I). A diagnostic tool for integrated analysis of complex problems and innovation capacity", *Agricultural Systems* 135, pp.1–11 (2015) https://www.sciencedirect.com/science/article/pii/S0308521X14001115

Further reading not on open access on the internet

8. Farmer First: Farmer *Innovation and Agricultural Research*, edited by Robert Chambers, Arnold Pacey and Lori Ann Thrupp. Intermediate Technology Publications, 1989.
9. Ian Scoones and John Thompson *Beyond Farmer First: New directions in pastoral development in Africa: Rural People's Knowledge, Agricultural Research and Extension Practice*. Practical Action Publishing, 1994.
10. Jens Andersson, "Potato Cultivation in the Uporoto Mountains, Tanzania: An analysis of the social nature of agro-technological change", *African Affairs* Vol. 95 (1996), pp.85–106.

11. B.M.L Namwata, J. Lwelamira, and O.B. Mzirai, "Adoption of improved agricultural technologies for Irish potatoes (*Solanum tuberosum*) among farmers in Mbeya Rural district, Tanzania: A case of Ilungu ward", *Journal of Animal & Plant Sciences*. Vol. 8, Issue 1 (2010) pp.927–935.

Topics for essays or exam questions

1. Consider an innovation made by small farmers whom you know. How did it spread? What could be done to make it possible for innovations such as this to spread faster and more widely?
2. Discuss the possible future contributions of social media to spreading new practices and ideas in rural Africa.
3. Consider the different words used to describe extension workers around the world. Which of these give the best impression of what the work involves? Do you agree that the term *Bwana Shamba* is sexist and top-down?
4. How could the advice given to extension workers be made more up to date? Should it be updated according to the seasons, and when the prices for crops change?
5. Extension workers often spend much of their time with larger farmers and with men rather than women. How can they give more assistance to smaller farmers and to women?
6. "Ultimately it is the knowledge already held by farmers which is the key to successful innovation in the future, and the purpose of extension should be to help this happen and then to spread the good news." How can this knowledge be understood and used?
7. Why has it proved so hard to achieve efficient production and processing of cashew nuts in Tanzania, when Vietnam managed to create a productive sector in just a few years?
8. An article entitled the "Rapid Appraisal of Agricultural Innovation Systems" [7] describes the effect of weeds on the growing of paddy in Tanzania. Why has there been little research on this? What research would you commission or undertake, to help farmers deal with the problem of weeds?
9. How can the research results be communicated to the farmers?
10. What advice should extension workers give to farmers to help them adapt to climate change?

PART 3

PRACTICE AND POLICIES

This section has just three chapters.
The first describes the issues raised by genetic
modification, and suggests alternatives which are less
environmentally and socially damaging.
The second chapter is about gender, and what is needed
to enable more women to develop their talents,
in agriculture and elsewhere.
The third is a digest of the agricultural policies arising
from the discussions in this book.

CHAPTER 10

Purchased Chemicals, Genetically Modified Seeds – and the Alternatives

Key themes or concepts discussed in this chapter

- Why the so-called Green Revolutions in the 1960s and 1970s in South America and Asia, which depended on heavy uses of fertilisers, insecticides, fungicides, weedkillers, and irrigation ceased to be effective in raising production.
- The unintended short-term consequences and dangers to humans of the use of purchased chemicals in agriculture.
- The long-term consequences when nature reacts to the repeated use of chemicals.
- The risks of losing biodiversity, and how this will be a loss to Africa and the world.
- Genetic modification and the issues it raises for the medium and long terms.
- Alternatives to the use of chemicals.
- How organic farming can succeed in tropical Africa, as it can elsewhere; but how it is likely to require more labour and produce less, though without the costs of many purchased inputs.
- The evidence that organic farming can produce good incomes, especially for small farmers.

THE GREEN REVOLUTIONS OF THE 1960S AND 1970S

In 1941 the Rockefeller Foundation and the Mexican government agreed on a programme to develop high-yielding varieties of food crops, especially wheat, which could be grown on large mechanised farms. A few years later, new varieties of maize were promoted in India, even though it was not an important food crop in that country, and new varieties of rice were developed in the Philippines.

These all depended on hybrid seeds and required heavy applications of chemical fertilisers, insecticides, and (where possible) supplementary water. The Rockefeller Foundation (and later also the Ford Foundation) worked closely with the international chemical and seed companies and concentrated their activities where there were large or relatively large mechanised farms with access to irrigation water.

What became known as the Green Revolution was further extended, for countries willing to allow the use of genetically modified seeds, when the company Monsanto created varieties of maize, lucerne and other crops that would not be affected by its weed killer $Roundup^{TM}$, which is based on the active ingredient glyphosate [1]. (See also Chapter 7, [3].)

However, by 1970 this first phase of the Green Revolution was winding down. It had led to greatly increased production of cereals, with very large areas planted with identical varieties. But new pests and weeds were appearing, yields were no longer increasing, much of the irrigated land was becoming saline, and the wider consequences were becoming recognised, including the fact that many small farmers had become heavily indebted and driven into poverty. The problems facing small and marginalised farmers were made even worse when they were unable to purchase the new seeds, pesticides and fertilisers.

There were also adverse public health consequences from use of the powerful insecticides and weed killers, as explained below. And governments were becoming reluctant to provide the subsidies for fertilisers on which the programmes depended.

But this was not the end. Research stations and seed companies were still creating new varieties, and there were parts of the world where the technologies had not spread widely, especially in Africa. The technology is scale-neutral, in the sense that both small and large farmers can use purchased seeds, fertilisers and sprays. But it is easier for suppliers to deal with a few large farmers rather than many small ones, and many small farmers do not have the resources or credit to purchase inputs.

There were, however, successes. Thus in 1986 Rasmussen wrote about "The Green Revolution in the Southern Highlands" in Tanzania, based on improved varieties of maize (not all hybrids) and subsidised fertilisers, grown mainly on small farms (Chapter 2, [8]). Since then hybrids have spread widely, enabling the south-west of the country to become the bread-basket of Tanzania.

A NEW GREEN REVOLUTION IN AFRICA?

Is a green revolution in Africa based on purchased chemicals the best way forward? The successful businessman and philanthropist Bill Gates thinks so, or he would not have called the institution he supports AGRA – Alliance for a Green Revolution in Africa. Many agronomists and economists think so: especially some who have studied the practice and the economics of farms in the USA and many parts of Europe which are becoming ever more dependent on purchased inputs and mechanisation. The companies and research scientists who have created genetically modified seeds think so, because most of their work depends on what amounts to another green revolution. President Kagame of Rwanda evidently thinks so, because, according to an article by Dawson et al [2], in parts of Rwanda he has forced large numbers of farmers to devote most of their land to a single crop, maize, grown with purchased fertilisers and insecticides, even when this means that less labour is required than previously, and many people have been forced out of agriculture. Approval is implicit, too, in many programmes supported by the World Bank, the FAO, and large multi-national companies.

This chapter shows that this approach is over-simplified [3]. There are times and places where it is appropriate to use purchased chemicals – indeed there are some situations where there is little alternative. But there are also downsides to their use: in the short term, and also in the long term if they are used repeatedly over many years.

The chapter also explains how genetic modification is carried out, and why altering genes can have unplanned consequences. Finally the chapter shows how organic farming – agriculture which does not use purchased chemical inputs (already introduced in Chapter 7) – is a feasible alternative in some situations, and considers its economic implications. The conclusion is not that purchased chemicals should never be used, but rather that they should be used sparingly, and only when there is no other alternative. In many situations farmers may be able to make a good living without them.

THE PROBLEMS AND RISKS OF INORGANIC FERTILISERS

The main chemicals needed for plants to grow were introduced in Chapter 1, where it was also pointed out that permanent cropping – using a plot of land continuously – is likely to lead to shortages of some naturally occurring chemicals, and hence to reduced yields.

For someone with money, the easy way to replace any chemical that is lacking in the soil is to add the appropriate purchased fertilisers. If fertiliser is not to be wasted, a farmer needs to know exactly which chemicals to add, and how much of each.

The alternative is to return nutrients to the soil through other plant or animal materials – as explained in Chapter 1 – through green manure or compost; plants such as legumes which "fix" nitrogen from the atmosphere; animal wastes; bone meal which is rich in phosphates and potassium; and materials that would otherwise be wasted, such as the husks and pulp which remain when coffee beans are separated from the berries. If, for any reason, crops such as tomatoes or mangos are not harvested, and are allowed to rot, these too will return nutrients to the soil. The difficulty with these alternatives is that large quantities of organic material have to be obtained (by growing them or transporting from elsewhere), and this involves high costs, particularly of labour and transport. Sufficient quantities may also be hard to obtain.

However, additions of inorganic fertiliser can increase the risks which farmers face. In many parts of Africa small farmers will use inorganic fertilisers when these are free, or available at a subsidised price, but not when they have to pay the full price for them, or take on credit obligations to obtain them – hence the prevalence of subsidies for fertilisers and other inputs. (The difficulties in persuading farmers to take up credit were considered in Chapter 8.)

Plants can only absorb a certain amount of fertiliser, so anything extra is a waste of money. Even worse, if excess nutrients are washed from fields by rain they may get into drinking water and make it unfit for human or animal consumption. Fertiliser which ends up in streams and lakes can cause very rapid growth of unwanted plants, such as the water hyacinth which is a problem on Lake Victoria and many other lakes. These plants may use all the oxygen in the water, reducing its capacity to support fish, and sometimes giving it a bad taste and making it unfit for drinking. Too much fertiliser can also poison plants or fish. It is far worse to use too much fertiliser than to use too little.

So farmers should be encouraged to use fertilisers only when they are needed: when it is known that a particular fertiliser will be effective, and when there are no easy alternatives.

INSECTICIDES, FUNGICIDES, ACARICIDES AND OTHER CHEMICALS

Information about the wide range of pests and diseases which attack agricultural crops in Africa was set out in Chapter 1. One way of trying to deal with them is to use manufactured chemicals. *Insecticides* are chemicals which kill insects. *Fungicides* kill fungi. *Antibiotics* kill bacteria. Acaricides are insecticides used to kill ticks and mites, which feed on the blood of animals.

Insects are not easy to kill, so the chemicals that can kill them have to be powerful – and therefore dangerous. In particular, these chemicals can cause long term damage to the nervous systems of animals and human beings. So great care needs to be taken to ensure that sprays do not drift on the wind anywhere near people or dwelling houses, and that the skin and lungs of the operators do not come into contact with the insecticides. The article by Mrema, Ngowi and others [4] suggests that women who work on horticultural farms are particularly at risk from exposure to fungicidal and insecticidal sprays.

Insecticides are usually applied by spraying. The chemicals are dissolved either in water or in an oil, and the mixture is sprayed to produce tiny droplets. For a large farm, a sprayer may be mounted on a tractor, and a wide area of crop can be sprayed in a single run. A very large farm may use a small aircraft or a helicopter! On a small farm, the sprayer may be carried on the back of an operator, and activated using a battery-powered pump or a hand pump.

Fungi are even harder to kill, and many chemicals which will kill them are of little use to farmers because they will also kill the plants that the farmers are trying to grow. There are some fungicides that are reasonably effective, such as copper sulphate to combat coffee berry disease, or sulphur dust to combat powdery mildew in cashew nuts. But for most fungal diseases some other method of control, such as burning or removing infected materials, may be the only possibility.

Ticks are very detrimental to animals. They carry diseases caused by viruses (such as anaplasmosis) and parasites (such as East Coast Fever). They cause itching, then sores, and finally bleeding which will restrict the growth of the animal. The normal method of control is by spraying or submerging the livestock in a bath or dip containing a chemical (acaricide) which kills the ticks. Once the acaricide has done its work, and an animal is "clean", it is again vulnerable to attack by other ticks. So, as far as possible, it needs to be kept away from animals which have not been treated, or from the grasses on which the ticks lie in wait. If the ticks return, the animal will need to be treated again.

But acaricides, either in dips in which the animals are submerged, or through spraying, usually by hand, are dangerous chemicals. So those who use them must take care not to breathe in drops from sprays, not to allow the chemicals to come into contact with their hands or faces, and to keep them away from drinking water. It is also dangerous to eat meat from animals too soon after they have been sprayed or dipped. [5]

Because all these chemicals are powerful, they kill not just a single pest, but many beneficial organisms as well, for example bees, wasps and butterflies that pollinate plants (and, in the case of bees, make honey). Insects that eat organic material and release its components into the soil for use by plants, and insects that eat other insects which are harmful to crops, are also likely to be killed by indiscriminate use of chemicals.

Over time, insects may become resistant to the insecticides. An insect that can survive an insecticide will have few competitors, so it will grow and multiply fast – and spread rapidly. Farmers then have to use ever-higher quantities of chemicals to control it. The chemical companies constantly produce new chemicals, trying to keep ahead of the emergence of resistant pests. The profits of the farmers go down. For obvious reasons, this problem of resistance is worse with monocultures, the same crop planted year after year on the same land. It is therefore very important to control infestations of insects and other pests *before* they develop resistance to the chemicals used against them.

ALTERNATIVES TO INORGANIC INSECTICIDES

The key to the problem is to control the extent and intensity of a pest infestation. It is often not necessary to eliminate entirely whatever is attacking the plants: a small amount of loss or disfiguration of the plant is usually acceptable. The problem will be less if plots are small, if crops are interplanted and not all planted at the same time, and if they are planted at times when the disease or pest is least active. These, as noted in Chapter 5, are all means by which small farmers reduce the risks they face. They also mean that there is less of the host of any particular predator or disease in a given area, so any attack will spread more slowly.

Some plants produce chemicals which kill insects. A good example is pyrethrum, grown as a cash crop in the Southern Highlands of Tanzania. The chemicals may be extracted and used in sprays. This is a plant product, so its use avoids most of the problems that can arise with inorganic insecticides. Similar chemicals ("synthetic pyrethrins") can be synthesised in factories. They are cheaper than the natural products, but they degrade more slowly, so they are more harmful to the environment. In recent years there has been a small revival in the use of the natural pyrethrum products.

An alternative to using chemical methods of controlling pests and diseases is to find or develop varieties of plants or animals that have natural resistance to pests or diseases, as discussed in Chapter 4. This takes time, and needs continuing work as new varieties of the pests or diseases are always likely to emerge.

In nature every organism has a predator – since otherwise they would multiply without limit. So killing those predators, through use of insecticides or any other practices, may well be a bad idea. But it may be possible to introduce new predators or diseases which will attack the cause of the problem. This is called "biological control". It is how the mealybug, a major pest which attacks cassava, was dealt with in Africa, with great success – as explained in Chapter 1.

It may be possible to control predators, or their effects, by hand. Thus if cotton is hand harvested, the workers can discard cotton bolls that have been stained by bollworms, and this will assist in getting a good quality product (and, in theory, a higher price). Birds and monkeys are often scared away from crops by people. Mice and rats can be controlled by cats. Diseased material, such as leaves, can be removed from crops and burnt. Caterpillars which feed on the leaves of vegetables such as cabbages or maize can be removed by hand – though this is impractical if the crop covers a large area. An extreme method of direct control is when the nests of *quelea* birds are bombed from the air.

There are plants, such as marigolds or onions, which deter certain insects. These may be planted close to crops which are threatened. Or they may be planted around a crop, to create a barrier.

THE LONG-TERM RISKS OF LOSING BIODIVERSITY

Over thousands of years, many different varieties of plants and animals have evolved naturally. Human beings have assisted in this process by selecting plants or animals that have the properties they prefer, and planting them or using them in breeding programmes. These varieties are adapted to local conditions. Many are resistant to drought. Others have properties of taste, smell, ease of storage, or cooking which make them attractive. In natural conditions the spread of any one variety or species is controlled, either by the climate (for example very hot or very cold periods of the year), or by other plants or animals. If it was not so, a species that was not controlled would take over the whole area.

In contrast, large international companies prefer to market a small number of crop varieties, which have uniform properties. These are usually bred to give maximum yields with high inputs of fertilisers and other chemical inputs such as sprays. Then the companies can multiply and sell the seeds, and the inputs which support them.

But if this takes place widely, many local varieties will die out. The varieties which remain may not be so drought resistant, so there will be increased risks of crop failure. They may require large quantities of water, as with some varieties of bananas and coconuts which have been released in East Africa, and which have not, in the end, been popular – for that very reason. These varieties may have poor resistance to insects which attack stored products. And while they may have high yields, they may not have the cooking or taste properties which many people remember and prefer.

If insecticides are widely used, many of the insects which feed on other insects will also disappear.

Biodiversity – maintaining a wide range of plant varieties and animal species – not only reduces risks and improves the quality of diets; it is also one of the ways of lessening dependence on chemicals. It should not be lost lightly.

THE MISUSE OF ANTIBIOTICS IN INTENSIVE LIVESTOCK PRODUCTION

There are similar problems with the over-use or misuse of antibiotic drugs in intensive livestock production.

Livestock may be raised in confined conditions where they are fed large quantities of food but can hardly move. Chickens kept like this may lose the use of their legs. This kind of confinement makes the animals vulnerable to diseases, which spread quickly. Such methods of livestock production require purchased inputs – manufactured animal feeds, and drugs to limit the spread of diseases. Furthermore, a category of drugs which are very important for human survival, antibiotics, are widely used – not just to control outbreaks of disease, but also to prevent bacterial infections, especially in intensively raised chickens and pigs, but also cows when they are kept in mega-dairies. More antibiotic drugs are used on animal farms than directly on human beings ([6], e.g. p.5). The conditions on such farms are very favourable to the appearance and spread of bacteria which are resistant to antibiotics. This is one of the reasons, perhaps the main reason, why, across the whole world, many antibiotics are no longer effective.

Animals and birds are healthier if they can move around freely, and their meat and other products often taste better. There are therefore regulations in most countries specifying minimum space requirements for animals, and premium prices are charged for "free range eggs" (eggs from chickens that have freedom to move around) and organic beef. Regulations governing the use of antibiotics are likely to be strengthened in the future, to prevent their misuse and over-use, and to ensure that the relatively small number of antibiotics that are recognised to be safe for treating human beings remain available as long as possible, for combatting diseases in humans.

GENETIC MODIFICATION

The arguments surrounding the use of artificial methods for raising crop or livestock yields become more contentious when "genetically modified organisms" (GMOs) or seeds are considered (see Fagan, Antoniou and Robinson [7].)

Living organisms are made up of cells – the smallest organisms have just one cell; complex organisms such as human beings have millions. Every cell contains within it a copy of the complete *genome*, or genetic information, from which the whole organism can be copied or created. Replicating this information is the hidden mechanism by which plants and animals produce "offspring".

The genome is made up of *genes*. Cells of human beings contain over 20,000 genes, but some agricultural crops have twice as many. So the number of genes is not important; what matters is what each gene does. Each gene – or more usually a group of genes working together – contains the information that produces the *traits*, or properties, of that organism, such as the colour of a person's eyes or whether a plant is resistant to a pest or disease. There are many more traits or functions of genes than there are genes; so most genes influence many traits or properties. It follows that changing a single gene can change many different traits of a given organism.

Each gene is composed of sequences of DNA. (DNA stands for deoxyribonucleic acid – but it is almost always referred to just as "DNA".) Most of this DNA consists of long strands of molecules, arranged in the form of a helix (shaped like a spiral staircase), which has the remarkable property of being able to copy itself in order to reproduce. As mentioned, parts of a gene, or group of genes, specify the traits of the organism;

other parts determine whether the organism retains this trait when it reproduces. This is how the DNA in a gene – and hence the whole gene and the traits and properties associated with it – is copied.

The process of copying DNA and genes is incredibly accurate. But very occasionally – on average no more than once or twice in the lifetime of a typical human being – a gene is produced which is slightly different from its original. This is called a *mutation*. Most mutations adversely affect the organisms concerned, so they do not survive. Other mutations are inconsequential, and stay in the new DNA without making a noticeable difference. But just occasionally a mutation gives a gene an advantage over others, and the mutated organism survives. It is through this process that organisms change and develop. That is how the huge diversity of plants and animal species on this planet emerged.

Once they had identified some of the genes in an organism, scientists started experimenting with them in what is now called *genetic engineering*. They found ways of introducing genes, or parts of genes, from one organism into the genome of a different species or variety.

Two particular genes have become very important in plant breeding. One is a gene that produces a chemical (or insecticide) which kills insects. The other is a gene which enables a plant to resist a particular herbicide. It is estimated that over 80% of all the genetically modified (GM) seeds used in the world incorporate one or other of these two genes.

The first of these, usually called Bt (its full name is *Bacillus thuringiensis*), came from a bacterium found in soils, which has the property of killing certain kinds of insects, especially aphids (which suck sap from plants), caterpillars and moths. Plants that contain this gene are used by organic farmers as a naturally occurring insecticide. The genetic engineers incorporated it into new seeds – and insects that eat leaves or suck juices from plants grown from these seeds, such as most genetically modified cotton, will die.

The second commonly introduced gene was discovered by accident in a laboratory, when scientists noticed that plants which included a certain enzyme were not destroyed by glyphosate, the main ingredient in the weedkiller *Roundup*™ developed and promoted by the chemical company, Monsanto.[14] The scientists identified the gene which gives the enzyme this trait and introduced it into "Roundup-resistant" varieties

---

[14] In 2018, the Monsanto company was taken over by the German pharmaceutical and chemical company Bayer. The name Monsanto will no longer be used, but the names of its products, such as *Roundup*™ will continue to be used.

of maize, soya beans, lucerne and cotton. The Roundup destroys the surrounding weeds, but not the genetically modified plants, which are very widely grown in the USA, Brazil and some other countries.

A third gene which has been inserted into a number of plant species gives them greater tolerance of drought. If this trait can be retained without having unwanted side effects, it will become extremely important in the context of the changes to the climate that are expected in the future.

The potential risks and benefits of reliance on genetically modified organisms

The greatest risk with GM is that unintended traits may be added to plants, with damaging consequences. These traits may take the form of harmful growth of cells – that is, cancers, which may not emerge for many years. Cancers can exist in humans for twenty years or more before becoming active. Most tests on plants for "toxicity" are carried out over much shorter periods, so the risk of adverse effects on those who eat them may easily be missed.

There is also a risk that modified genes will spread to plants in surrounding areas. They may even spread to weeds, through natural processes of cross pollination. If this happens, these weeds may become resistant to the weedkillers and there will be nothing to stop them spreading. Then either the farmers will have to spend more money on weeding or their yields will decline.

An additional side-effect of planting genetically modified crops is the potential increase in the need for pesticides and weedkillers. We have already described the harmful effects of their possible residue in food and water supplies, and of the elimination of beneficial or useful insects.

Last but not least, it is likely that the short term benefits brought by genetically modified varieties like Bt cotton or Roundup-resistant soya may not be sustained over the longer term, as insect pests or weeds will emerge that are not killed by the chemicals they produce. These resistant pests and weeds will have little competition, so they will spread fast, and ever-increasing doses of chemicals will be needed to control them.

This appears to be why, in both the United States and India, the new varieties gave very good results in the short term, but in the long term have not proved to be sustainable. For these and other reasons, genetically modified seeds remain banned in Europe (with some very specialised exceptions), and in most countries in Africa they are being used only in carefully controlled research trials.

However, it is important that the downsides of GMOs are not exaggerated, any more than the upsides. Some genes have been modified with clearly beneficial effects, at least in the short term. Thus Water Efficient Maize for Africa (WEMA) – a variety of maize genetically modified for greater drought tolerance – requires less water and survives drought better than most other varieties. It is being trialled in Kenya, Mozambique, South Africa, Tanzania and Uganda. Genes that increase protein content have been added to some cereals and to cassava. Genes that enable plants to tolerate a high salt content of soil or water have been added to rice. The direct risks of the GM cereals and legume crops already released in the United States and large parts of South America must be very low indeed, since the whole population of these countries has been exposed to them for many years now, without apparent problems.

Nevertheless, the long-term threats to biodiversity and to the spread of resistance remain, as do the risks of monopoly control of the supply of improved seeds and agrochemicals by a small number of multinational companies. This last threat will be controlled to some extent if more seed and agrochemical companies enter the market, and more new varieties and products are developed. This would ensure competition, from which the users of the new seeds and chemicals would benefit.

The best way forward for Africa in the GMO debate remains unclear, and in these circumstances individual countries are wisely moving forward with caution.

THE ECONOMICS OF ORGANIC FARMING

The extensive use of chemicals and of large machines, is not the only way of using land productively, though many feel that it is the only way to feed rapidly expanding world populations. Much of the new technology has been developed to minimise the use of labour in countries where labour is expensive (though it does not entirely succeed, because large labour forces are still needed to harvest many fruits and vegetables, if only for a few weeks or months in a year – often this is cheap migrant labour from other countries). Much of the value added goes to the suppliers of the chemicals, the seeds, and the machines.

The opposite extreme to dependence on purchased chemicals is *organic farming*, which means using no purchased chemicals at all [8] [9].

Almost inevitably, organic farming requires additional labour, and this adds to costs if the labour is paid. (If the labour is supplied by family members then there may be no extra labour costs.) For example, the application of compost or green manure to a crop needs far more labour than the mechanical application of an inorganic fertiliser. If a "barrier crop" has to be planted to protect a farmer's crop from insect attacks, then productive land is used for a crop that gives only a small return. If seeds are kept by the farmer from previous years, the crop is unlikely to be as uniform as it would be if the seeds were purchased anew. If a farmer applies for organic accreditation in order to get higher prices, as discussed in Chapter 7, then the costs of this accreditation and ongoing inspections have to be paid.

Moreover, yields with organic farming are likely to be lower, but of course the costs of inputs will also be lower. So the overall profitability depends on whether the higher prices obtained compensate for the lower yields, and on how the family's labour is costed. If the family has surplus labour, so that the members would not earn much by working elsewhere, then it is reasonable to cost that labour at zero, or a very low figure. Under these circumstances the lower yields may be more than compensated for by the savings on purchased inputs. This was the conclusion of trials carried out on organic farms in the Uluguru mountains near Morogoro in Tanzania by Chie Miyashita [10].

"Conservation agriculture" (or "no-till agriculture", or "minimum tillage"), as explained in Chapter 2, falls between the two extremes of organic agriculture and agriculture based on high levels of purchased inputs. It uses a minimal amount of machinery, avoids opening up the soil to water loss and erosion, and maximises retention of organic matter for improving soil structure and fertility. This is good practice, which can be of benefit even to a farm with huge fields dependent on purchased chemicals.

CONCLUSIONS

The Places of "Chemical" and Organic Farming

There is no doubt that farming which is dependent on purchased chemicals poses many threats to the environment, and to public health. By promoting products which are uniform, often from exactly similar seeds used over very large areas, chemical inputs increase risks to the environment and threaten biodiversity.

On the other hand there are situations and crops where agriculture is very difficult without chemical inputs, and where the use of hybrid seeds and even GM crops may become unavoidable in future, despite the recognised risks, if the growing population is to be fed.

Where organic alternatives are available, whether they are used will depend on:

- the yields that can be achieved, and how these compare with the yields using inorganic fertilisers and agrochemicals;
- whether the organically produced crop will attract premium prices – as organic products do in at least some export markets;
- how the extra labour needed is costed. If it has to be paid at a commercial rate, the farm may not be profitable; if it is paid little, because undertaken by members of the farmer's family, then organic farming may be profitable.

Organic farming will also be less risky, and bring more benefits in terms of sustainability, for the whole environment of the area.

Regardless of the above, conservation agriculture should be supported wherever it is possible, and especially where soils are weak and at risk of erosion.

The debates surrounding genetically modified crops will continue around the world for many years to come. There could still be long-term adverse consequences which have not yet emerged. But more serious in the short term is the dependence on a relatively small number of seed varieties, and the risks that a few large companies will use them to increase sales of their insecticides and weedkillers – all of which are very dangerous chemicals which can easily harm those who use them or live near to where they are used.

Case Study 14: Data on the Profitability of Organic Farms

This case study is based on the 2015 Masters degree dissertation by Chie Miyashita: "Can organic farming be an alternative to improve well-being of smallholder farmers in disadvantaged areas? A case study of Morogoro Region, Tanzania" ([10], and see also [11]).

Miyashita compared two samples of farmers: 160 farmers from 20 villages in the Uluguru mountains near Morogoro, selected because they were carrying out agro-ecological practices, and a comparison sample of 164 farmers randomly selected from four villages not far away. Of the agroecological farmers, 90% were using organic fertilisers, 72% organic pesticides, and 81% mulching. The comparison farmers' figures for these practices were far lower (Table 10.1).

Table 10.1: Farming practices

Farming practices	Agro-ecological farmers (N = 160) (%)	Comparison farmers (N = 164) (%)
Organic fertilisers	90.0	35.4
Organic pesticides	71.9	3.0
Crop rotation	81.9	25.6
Mulching	81.2	19.5
Terracing	63.1	34.1
Intercropping	75.0	74.4
Cover crops	88.8	78.0
Chemical fertilisers	13.8	21.3
Chemical pesticides	13.8	21.3

The agro-ecological farmers had much better marketing arrangements than the comparison sample. They had access to reliable markets, went to markets much more often, and had regular customers (Table 10.2). They achieved far higher prices (Table 10.3): 50% higher for rice and bananas, and significantly higher for maize. (Note: the numbers of farmers in this table are low, because many farmers kept most of the maize and other crops they grew for food.) For Chinese cabbage, a somewhat specialised product, their access to good marketing outlets enabled the agro-ecological farmers to get three times the price of the comparison farmers. However for cowpeas and pumpkins, the comparison farmers got higher prices than the agro-ecological farmers. Table 10.4 shows that, on average, yields did not differ substantially, other than for pumpkins, where the agro-ecological farmers got higher yields.

Table 10.5 shows that the agro-ecological farmers got far higher incomes – nearly five times as much on average – though one managed to make a loss. One of the agro-ecological farmers made a huge income, so there must be something special about this farm – it might have been better not to include it in the sample.

Table 10.2: Market access

Market engagement	Response	Agro-ecological farmers (N=160)	Comparison farmers (N = 164)
Do they have a reliable market?	Yes	69 (43.1%)	14 (8.5%)
Do they have a contract with trader/buyers?	Yes	12 (7.5%)	0 (0%)
How often do they go to a market?	At least once a week	51 (75.0%)	8 (57.1%)
	At least once a month	14 (20.6%)	1 (7.1%)
	Less than once a month	3 (4.4%)	5 (35.7%)
Do they have regular customers?	Yes	71 (44.4%)	13 (7.9%)

Table 10.3: Average price of 1kg of crop products sold in 2013 (TZ Shillings)

Crop	Agro-ecological farmers				Comparison farmers			
	N*	Min	Max	Average	N*	Min	Max	Average
Maize	31	150	1400	526	40	30	750	490
Rice	16	100	1500	745	3	400	700	530
Banana	95	80	1080	304	7	100	450	202
Cow pea	13	70	2910	920	19	320	2000	1071
Pumpkin	25	114	700	305	19	100	1800	516
Chinese cabbage	53	400	2000	1242	8	50	1200	407
Tomato	11	150	1000	498	13	125	1800	490
Cabbage	28	70	1600	342	80	100	666	214

* N is the number of farmers selling the product.

Table 10.4: Mean of estimated yield from 1 ha (kg)

Crop	Agro-ecological farmers		Comparison farmers	
	N*	Mean	N*	Mean
Maize	141	1156	162	1039
Cow pea	80	208	92	186
Pumpkin	95	410	81	261

* N is the number of farmers selling the product.

Table 10.5: Gross income, costs and profit in 2013
    (TZ. Shillings '000)

	Agro-ecological farmers (N = 160)			Comparison farmers (N = 164)		
	Min	Max	Mean	Min	Max	Mean
Gross income	0	56,005	1,843	0	1,726	379
Total costs	0	1,269	206	0	383	232
Profit	-391	54,736	1,637	-1,879	16,626	147

It is important to note that the comparisons are between two field samples, not between a sample and an ideal achieved on a research station. On a research station, yields would have been substantially higher all round. What the figures show is that, for these particular samples, most of the farmers who had adopted agro-ecological practices achieved very good prices and got very good incomes.

Further reading on open access on the internet

1.   Raj Patel. "The Long Green Revolution", *The Journal of Peasant Studies*,
      Vol. 40, no.1, pp.1–63 2003.
      http://dx.doi.org/10.1080/03066150.2012.719224
2.   Neil Dawson, Adrian Martin and Thomas Sikor "Green Revolution
      in sub-Saharan Africa: Implications of Imposed Innovation for the
      Wellbeing of Rural Smallholders". *World Development* Vol. 78, pp.204–
      218, 2016
      http://www.sciencedirect.com/science/article/pii/S0305750X15002302
3.   Jules Pretty: "Agroecology in Developing Countries: The Promise of a
      Sustainable Harvest", *Environment: Science and Policy for Sustainable
      Development*, Vol. 45 No.9, pp.8–20 2003. https://www.usc-canada.org/
      UserFiles/File/agroecology-in-developing-countries.pdf
4.   Ezra J Mrema, A V Ngowi, S Kishinhi and S Mamuya "Pesticide
      Exposure and Health Problems Among Female Horticulture Workers in
      Tanzania" *Environmental Health Insights*, Volume 11 pp.1–13 2017. This
      article explains how women workers are the most at risk from exposure
      to pesticides in Tanzania. http://insights.sagepub.com/pesticide-
      exposure-and-health-problems-among-female-horticulture-worke-
      article-a6430
5.   "Experiences in Tick Control by Acaricide in the Traditional Cattle
      Sector in Zambia and Burkina Faso: Possible Environmental and Public
      Health Implications". By Daniele De Meneghi, F Stachurski and H Adakal
      (2016) *Frontiers in Public Health* Vol.4, Article 239, 2016.
      https://www.ncbi.nlm.nih.gov/pmc/articles/PMC5101216/
6.   *Antimicrobials in Agriculture and the Environment: Reducing Unnecessary
      Use and Waste.* Report of the Independent Review on Antimicrobial
      Resistance (Chair Jim O'Neill), London: Welcome Trust, 2015.
      https://www.nhs.uk/news/medication/antibiotic-use-in-farm-animals-
      threatens-human-health/
7.   John Fagan, Michael Antoniou, and Claire Robinson. GMO Myths and
      Truths: An evidence-based examination of the claims made for the safety
      and efficacy of genetically modified crops and foods. London: Earth
      Open Source, 2nd edn., 2014. www.earthopensource.org/wordpress/
      downloads/GMO-Myths-and-Truths-edition2.pdf
8.   IFOAM (2013) *Productivity and Profitability of Organic Farms Systems
      in East Africa*, International Federation of Organic Agricultural
      Movements. http://www.ifoam.bio/sites/default/files/page/files/osea_ii_
      oa_prod_prof_report_final.pdf
9.   Isgren, Eleanor and Barry Ness "Agroecology to Promote Just
      Sustainability Transitions: Analysis of a Civil Society Network in the
      Rwenzori Region, Western Uganda", *Sustainability* Vol.9 No.8 2017.
      http://www.mdpi.com/2071-1050/9/8/1357/htm

10. Miyashita, C. *Can organic farming be an alternative to improve well-being of smallholder farmers in disadvantaged areas? A case study of Morogoro Region, Tanzania,* Unpublished MA dissertation in Rural Development, Sokoine University of Agriculture. Morogoro, Tanzania, 2015 http://www.suaire.suanet.ac.tz:8080/xmlui/bitstream/ handle/123456789/1191/CHIE%20MIYASHITA. pdf?sequence=1&isAllowed=y

11. Anna *Mdee*, Alex *Wostry*, Andrew *Coulson* & Janet *Maro. The Potential for Inclusive Green Agricultural Transformation: Creating Sustainable Livelihood through an Agroecological Approach in Tanzania.* Chronic Poverty Advisory Network, 2017. https://dl.orangedox.com/Tzm5EF

Topics for essays or exam questions

1. Discuss why the Green Revolutions in South America and Asia ceased to produce increases in productivity after a number of years. What implications can you draw from this for Africa?

2. Discuss the reasons why small farmers are often reluctant to use inorganic fertilisers, and set out the advantages and disadvantages of the alternative ways in which they can put nutrients back into the soil.

3. Discuss the advantages and disadvantages of chemical insecticides and herbicides. Explain why they often lose their effectiveness if they are used on the same fields over long periods of time. What are the main alternative methods of controlling insect or weed infestations?

4. Why is it important to preserve biodiversity? How can this best be achieved?

5. Intensive livestock production gives very high yields in the short term. What are the long-term disadvantages?

6. Summarise the arguments for and against the adoption of genetically modified organisms in Africa.

7. Summarise the arguments for and against use of the weed-killer Roundup™ in Africa.

8. Discuss the arguments for and against organic farming. What difficulties face small farmers who want to "go organic"?

9. Discuss the economics of organic farming. Under what circumstances will it be an attractive option for small farmers in Africa?

CHAPTER 11

Gender Myths and Half-Truths

Key themes or concepts discussed in this chapter

- This chapter is about why many rural women stay poor, and how they can confront the forces which make it so.
- There are many myths or half-truths about women and development. One of these is that women do more work in the fields than men. The latest studies suggest that men and women do about the same amount.
- Another myth is that men and women cultivate separate fields, and that men are much more productive than women – so there would be great increases in production if women could reach the same levels of production as men.
- Another myth is that women who are heads of farming households are all poor. Some are far from poor – and some are better off than they would be if they had to share what they produce with a less successful man.
- Poverty is found in many households headed by men – this is not a myth.
- There are issues which affect women which go beyond agriculture – gender-based violence, forms of discrimination including many in the law and in the way it works, and in access to finance and credit.
- Agricultural production will be improved by investments which save women's time, such as providing sources of clean water near where they live, more efficient cooking stoves that use less fuel, feeder roads and bridges to get crops to markets, and homes and equipment which are easier to manage.

TEN MYTHS OR HALF-TRUTHS ABOUT WOMEN AND DEVELOPMENT

This chapter examines a number of myths, or half-truths, which dominate much of the discourse about women and development in rural Africa. It builds on that discussion to show what a society needs to do if wants to improve the position of its female members – and hence of all of society. It draws, in particular, on the Introduction to a special issue of a journal which included papers from a conference on gender and development in 2003 [10].

Gender myth or half-truth 1: Most of the work in the fields is done by women

Until recently it was commonly stated that in most parts of sub-Saharan Africa women do much more work in the fields than men. The main source of this was a much quoted World Bank "Status Report" published in 1998 on poverty in Africa [1].

A more recent World Bank paper from 2015 [2] tells us that women do just over half the work in the fields. They do less where Muslim influence is strong, and where livestock and pastoralism are the main agricultural activities. Thus in the North of Nigeria women do less than half the work in the fields, but in the South they do more.

The figures are still contested (see interesting points made by Cheryl Doss [3]), and they vary from place to place. The 2015 figures are for work in the fields only, and do not include most of the work relating to livestock, which can involve collecting fodder, looking after small animals around the homestead, or herding cattle or goats. Nor do they include post-production work such as drying sunflower, sesame seeds, or cassava. Women's work as labourers on larger farms is also excluded.

It is not entirely clear how the amount of work done relates to seasons in the year. At certain times of the year, such as when a harvest needs to be collected, or crops planted while there is still moisture in the soil, it would be expected that every able-bodied person in a household would work long hours. At some other times, for example during long dry seasons, there is less pressure to work.

Traditionally, in most rural societies in Africa, women have taken primary responsibility for growing the food to feed the family and then processing, storing and cooking it, as well as looking after small animals such as chickens, sheep and goats, and dairy cows where these are kept in stalls in the homestead. Men were responsible for crops grown specifically for sale, large herds of cattle, and the buying and selling of livestock. Thus men were involved in most of the money-earning activities, while women were mainly involved in unpaid activities.

Even then the division of labour was not straightforward, because women assisted with many labour-intensive tasks on all crops, such as harvesting cotton or transplanting paddy. However, as we have seen, for many families today cash is earned by selling crops which are also grown for home consumption – cereals, fruits, vegetables, and legumes. Sometimes there are separate plots for food intended to be kept and stored, and food grown to be sold, but the distinction between men's crops and women's crops is breaking down.

Women work long hours in total, because they have many other responsibilities in addition to farming: childcare, cooking, cleaning, washing clothes, collecting firewood and water (men and children also help with these) and also often carrying crops from the fields to their homes or to markets. Women look after sick or elderly family members, manage the food stored in the homestead, choose good seeds after the harvest and carefully store them for planting in the next season, and earn money in many other ways. Given all this, it is an achievement that they have the time to do so much work in the fields.

The 1998 Status Report quotes studies in Tanzania and Ghana which found that household members spent between 1,150 and 1,490 hours a year – an average of 3 to 4 hours a day – moving loads on their heads – water, firewood and crops for sale or for cooking. Women transported about four times as much as men ([1], p.5).

So perhaps women do not spend more time in the fields than men. But they work very long hours on all their other commitments and still manage to spend long hours in the fields.

Gender half-truth 2: Men have more leisure time than women

This follows from Half-truth 1. The 1998 Status Report presents figures for a set of African countries. Thus in Uganda, women are reported to work on average 15 hours a day, while men work only 8.5 hours. In Kenya women are reported working for 12 hours a day, men for just over 8. In Tanzania, women are reported working for nearly 10 hours a day, men for nearly 8 ([1], p.4).

Again, there are uncertainties about the data. The Status Report tends to take a study of a few villages and from those villages to generalise about the whole country. Thus it uses a study by Anna Tibaijuka [11] which reported that women in the Kagera area of Tanzania work very long hours. It does not follow, though, that women in other parts of

Tanzania do the same. The differences between countries are too great to be convincing: it would be interesting to break them down by regions of countries, if the data existed.

The image of women working flat out, while men just laze around, cannot be the full story. It does not take account of the tasks that men do, at different times in the agricultural year – they may work long hours at peak periods, preparing land for planting, or weeding, and harvesting, and less during long dry seasons. This topic requires more research, starting with village studies of the demands for labour at different times in the year. But the question behind the myth remains: could men in many places work longer hours?

The confusion is added to by the third half-truth:

Gender half-truth 3: Men and women have separate plots, and the men cultivate their plots better than the women

The World Bank analysts of the 1990s were influenced by studies which found that women worked on their own plots, growing food for their families, while men worked on separate plots, growing crops for sale. The women used less fertiliser and insecticides and fewer improved seeds, and their yields were lower than those from plots cultivated by men, even when they were growing the same crops. From this the Status Report concluded that there would be an increase in production, and a reduction in poverty, if women used more fertilisers and other inputs. Katrine Saito's 1994 paper [4] is an example of an extensive literature that draws this conclusion. However, the more recent 2017 analysis of the methodologies of such studies by Cheryl Doss [5] shows that this is a very unsafe conclusion when all the issues are taken into account, especially the fact that, more and more commonly, women and men are working together on their farms.

In Kenya, the analysts argued that if women used the same quantities of agricultural inputs, and had the same education as men, their yields would be higher by more than 20 per cent ([1], p.20). So their policy recommendation suggested that production would increase, and poverty would be reduced, if women could be persuaded to adopt new technologies. (They did not say it, but a similar result would therefore arise if some of the plots were transferred from women to men!)

But, as argued in previous chapters, there may be good reasons for not using some of these technologies, especially on food crops which are vital to the survival of the family. For example, traditional local varieties may not respond well to fertilisers, but they may resist drought better than the new varieties. They may have desirable cooking or eating properties, and be good for storage. And extra purchases, or use of credit for buying inputs, increases the risks if for any reason a crop fails.

As already noted, there is increasing evidence that the rigid division of plots between women and men is breaking down. Men, not just women, are working on plots growing food for local consumption, especially with some of the heaviest tasks, such as preparing the land. This is especially likely where the foods grown for sale and for local consumption are the same (for example maize or rice). On many farms, crops are interplanted, and if there are surpluses, sales are made of any of the crops grown. If the cropping is planned and implemented jointly by men and women, and the income from any sales is shared, then the basic premise of the argument about improvement is wrong.

The Farming Systems studies discussed in Chapter 5 did not distinguish between men's and women's labour availabilities. Instead they looked at the total amounts of labour available to the family, and the specific tasks that must be done in each month of the year for each crop grown. This kind of data almost always shows that there are some months in the year when the amount of labour available (to the whole family, i.e. men and women together) limits how much can be grown – for example when land has to be prepared quickly to catch the rains, or weeded, or when a labour intensive crop such as cotton has to be harvested.

These studies also show that there is "leisure time" at certain periods of the year, as in long dry seasons, when little can be done in the fields. But that does not mean that the time is used for sitting around and drinking beer. There are other tasks, such as maintenance work on houses, tracks, furrows, or terraces. Water and firewood may have to be collected from further and further away. And "spare" time can often be used for making bricks, baskets, pots, or craft items for sale. Or becoming involved in small scale mining, or even travelling far away to find work.

Gender Half-truth 4: Women grow crops for food, men grow crops for cash and keep the money

This has been discussed above. There is some truth in it. But not all men keep all the money, and many women earn money by selling agricultural products which are also foods.

Gender Half-truth 5: Households headed by women are poor

The most common way in which a woman may become the head of a household is through widowhood – not least as a result of HIV/AIDS. Other women have fled from gender-related violence. Still others have been abandoned by men who have gone to work elsewhere. Some have inherited land, often very small plots.

Many female-headed households are poor, some very poor, as Da Corta and Magongo show for women in parts of southern Tanzania [6]. They face discrimination –for example in getting access to good land or irrigation water – and attempts are sometimes made to exploit or cheat them when they sell crops. But on average, as the Status Report [1] points out, families with households headed by women are no poorer than households taken as a whole. The skills involved in running a household are similar to those needed in business: deciding how much time to spend on each task, in what order to do it, whether to employ someone else to help – and living with the consequences if things do not work out as anticipated. So women heads of households who have access to sufficient good quality land, and some education, are often extremely successful. Some may be better off because they do not have to look after lazy men![15]

Conversely, women can be abused or neglected in households headed by men, and many of the poorest women are in such households. Poverty is not just a matter of gender, or of women not being good farmers. It is demeaning to blame women (or indeed men) for being poor without investigating the detailed circumstances of each case.

Gender Half-truth 6: Women are poor because they are discriminated against in their access to land and livestock

This is an over-simplification, since the law in most African societies gives women equal or near-equal rights. Even so, in practice women may be discriminated against in the access they have to land (there are exceptions, however, in matriarchal societies where the heads of households who hold the property are women). For example plots are often allocated to women by the head of the household, usually a man. But if that man dies, the customary laws of inheritance may prevent women from inheriting clan (or village) land, so women who have only female children are likely to lose their land. There have been

---

[15] One woman farmer, a widow with four children, growing sugarcane on an irrigated sugarcane project in Kenya, replied when asked how she managed without her husband, "Sugarcane built my house; sugarcane educated my children; sugarcane is my husband. What do I need of a man?"

improvements over time in the national laws, but the customary laws and traditional practices of inheritance remain dominant in the rural areas, where women generally have access to land only through marriage.

Livestock are important assets. They are a form of wealth, saving, security and often they also comprise part of the marriage payments. It is unusual in pastoralist societies for women to be involved in the buying and selling of livestock. The income that is earned from cattle is usually controlled by men.

Gender Half-truth 7: The problems will be solved when women have more access to education

These days it is not easy to be a good farmer without being able to read, to measure and to count. Much information is available on mobile phones, even more on the internet. Those who have little ability with numbers are at risk of being cheated by traders. Those who cannot use M-pesa and read information sheets are at a disadvantage. On the other hand, much of the important knowledge about agriculture and conservation is (or was) passed on verbally, from fathers to sons, from mothers to daughters. Most of this information – about how to cultivate, how to survive droughts and famines, which plants have medicinal properties for different illnesses or conditions, how to maintain the fertility of the soils – is not taught in schools, and cannot be learnt from books. There is clearly a need for education: as a minimum to be able to count money, calculate change, and read basic information. But there is also a need to maintain traditional forms of induction and initiation through which important skills are transferred.

This half-truth is sometimes used as an argument for family planning. Thus, statistically, the more education women have, the fewer children they are likely to have. And if there are fewer children, there are fewer dependent mouths to feed, and more time for "productive" activities such as agriculture. So a country that wants to slow its rate of population growth should ensure that its young women go to school. It is true that rapid population increase will make it more difficult for people to escape from poverty, especially families with many children. But to make reducing population growth a main argument for education is to underplay its importance in opening up other ways of getting information, and opening doors to many of the good things in life.

The quality of education matters as much as the quantity. If teachers are poorly trained, or their knowledge is out of date, education may do as much harm as good. If schools teach in a manner which implies that all knowledge and information is held at the top of an organisation and has to be passed down, they may not give their students confidence to make their own judgements and decisions. Basic literacy and numeracy are becoming increasingly important, but in modern life and business, independent thinking and judgement are essential skills. Women need good quality education, not just any education.

Gender Half-truth 8: The problems will be solved if extension workers orient their work to women

Studies have shown that male extension workers spend most of their time talking to men. They also spend much of their time with the better off, more successful, farmers, most of whom are men. It is obviously valuable to have extension workers who are female, as they find it easier than men to deal with women farmers. It is a welcome fact that more women are being trained and employed as extension workers.

But, as the previous chapter of this book explained, when farmers innovate, many of the new ideas do not come from extension workers, and most extension workers (of either sex) are not up to date in their knowledge, and are unable to adapt their recommendations in the light of changing market prices and opportunities. Many lack transport, and are poorly supervised – so much so that the chapter ended by questioning the value of village-based generalist extension workers. There are so few extension workers that most farmers will meet them very infrequently, or not at all. It is good to have more women extension workers, but unlikely that this in itself can transform the position of women in rural Africa.

Gender Half-truth 9: The problems will be solved if there are more women in top jobs

To use all its talents, any society needs the best people, regardless of gender, in all kinds of employment. A few jobs require physical strength, which men generally have more than women. Even then, there will be some women who can do these jobs. Having women active and visible in the top jobs, in politics, public administration, large companies, regulatory bodies, etc., provides good role models for younger women.

Of course it does not follow that women in top jobs are always good at them, or supportive of other women. Many women politicians who have got to the top have done so in a world dominated by men, and are comfortable in it. But their ideas and policies have not always helped the poor, women or men.

So this is another half-truth. Yes, more women are needed in jobs at the top – but not women who are out of touch with the realities of survival in harsh situations.

Gender Half-truth 10: Mainstreaming gender in public policy will solve the problems

In practice, mainstreaming gender in public policy often means having check lists of policies, and performance indicators that measure whether these policies are succeeding.

There is no doubt that targets can motivate public servants to change their behaviour, but they will not always change in the ways that those who created the targets expect. Thus a target to increase the number of children in secondary school may lead to a rapid expansion of secondary education, but if there are too few teachers and not enough books, and the teachers are not paid on time, and the syllabuses are out of date and lacking in imagination, not much is achieved. In a similar way, it is good to have targets for health, but there is also a need for drugs and equipment.

So targets may be met but the outcomes may disappoint, because those in charge have responded to the targets but not to the underlying situations that they were created to deal with. The real need is for people to understand *why* the targets are being set, and to have policies that make sure that achievements are real. That needs much more than a list of publicly quoted targets. It requires leadership which will ensure that those who want change are able to achieve it – a recognition of the realities of power and influence in any society.

IMPROVING THE POSITION OF WOMEN

All over the world, women are campaigning for greater fairness, freedom from oppression, and opportunities to use their talents. What is discussed here is a tiny part of a much bigger picture. There are implications at every level of society – from the way a country is governed, to how an organisation runs itself, down to what happens in households or families. Women are asking for changes in culture – in what is recognised as acceptable practice and what is not.

There has been clear progress. But a change in national culture may be more influenced not so much by government policies as by what people hear or see in films, or TV or radio "soap operas" with story-lines that tackle fundamental issues in society, such as domestic violence or discrimination against people with any kind of disability or difference.

Much of the argument of this book is supportive of traditional practices and recognises that a huge amount of relevant knowledge about the natural environment is passed from generation to generation. Societies will be much worse off if this knowledge is lost. Some of this knowledge, and the tasks that go with it, are gender specific, or have been up to now. But it is also true that there are flaws in many traditional practices, especially when these relate to the position of women. Thus it is not acceptable that some women should suffer domestic violence and have no way of escaping from it, or that women should have so many domestic tasks that they have to work much longer hours than men, or that women should not have access to good quality farm land. Nor is it acceptable that fewer girls than boys should go to school, or that there is discrimination against women in the allocation of water for irrigation, or in the money they get when they sell crops.

It suits certain people to have other people available who will work for low returns or wages. It also suits some large employers to pay women less than men for the same work, or not to promote them.

Very low wages or incomes for women, and poor prospects for the future, are not good for society as a whole. The talents of capable women are not being fully used, and half the working population feel exploited, disregarded and taken for granted. The situation may be improving, but slowly. To do it faster requires action by politicians, or governments, to make sure that talents are developed, used, and appreciated in the country as a whole, and at every level in every organisation. It also requires a willingness to discuss traditional practices, in different places, so that good practices can be preserved and enhanced, while ways are found to end bad practices that are damaging to women and hence to society as a whole.

In an organisation such as an NGO, or in an arm of bureaucracy such as a regional office, it is important that women are employed, and visible, at every level. Not all these women will be sympathetic to other women, but most will be. On a farm, of any size, men and women need to work, and work together as far as possible (as they now frequently do). They need also to plan together how any incomes will be spent, and how food is to be provided. That is the best way to ensure that all the talents are used. Women must also be encouraged to form their own organisations, such as groups of women farmers, or societies which represent the interests of women in large organisations (even though better off and more successful women may dominate such organisations).

The law, and the justice system, and its working in practice, needs to be kept under review to ensure that it does not discriminate against women, in, for example, land law, inheritance law, or employment law. Penalties for those who abuse or discriminate against women, and the positions of those who speak out against injustices of any kind should be respected and protected.

Often it is indirect investments that make the most difference to women. For example, investment in reliable, clean water supplies for household use within half a kilometre can save hours of time carrying water. That time can, in principle, be used in the fields or for other economic activities. Improved cooking stoves may save fuel, and also lessen the chances of fire, and the health risks from breathing in smoke. Investment in feeder roads, tracks, and bridges may make it possible for pick-ups or oxcarts to reach the fields, and for crops to be transported safely and easily to markets, without so much carrying on people's heads. Stores which can be fumigated safely may enable crops to be kept for several months, and possibly sold later at higher prices.

The article by Koopman and Faye [7] shows that when there is land-grabbing – a company or a big farmer trying to take land away from small farmers – women are active in fighting it. Why, they ask, are women not successful in organising themselves to campaign against the many other injustices that they face?

Finally, here is a quotation from a paper which summarised the conclusions of a conference on Gender and Development in 2003 [10]:

> Lessons learnt from particular places have been turned into sloganised generalities: "women are the poorest of the poor", "women do most of the work in African agriculture", "educating girls leads to economic development" ... and so on. Some have been used as Trojan Horses to open up debates and advocate positions. Others have become popular preconceptions, useful as a kind of catchy shorthand to capture the policy limelight. Others take the shape of feminist fables, cautionary tales told with educative intent. And still others gain the status of myths, stories whose potency rests in their resonance with deep-rooted convictions. Women appear ... as abject victims, the passive subject of development's rescue, and splendid heroines, whose unsung virtues and whose contributions to development need to be heeded.

Case Study 15: Labour Allocation of Women, Men and Children in Tanzania

This case study is taken from the 2009 PhD thesis of Romanus Dimoso on how environmental degradation affects the allocation of labour between men, women and children [8].

Dimoso undertook fieldwork in the South Pare Mountains of Tanzania, where population pressure has led to degradation of the soils, erosion, and loss of forests which provided firewood as well as protecting water sources. He studied ten villages. Three were in the highlands, where degradation of the environment was less of an issue. Three were in the foothills, or "middle plateau" area – where degradation is an issue. Four villages were in the lowlands where degradation is a very severe issue indeed.

Here are some of his figures:

Table 11.1: Mean hours per day spent on farm work, fuel wood and water collection in ten villages in the South Pare Highlands, Tanzania

	Non-degraded (the upland plateau zone)			Medium-degraded (the middle plateau zone)			Severely-degraded (the lowland zone)			
	Gwa-nga	Gonz-anja	Kira-ngare	Mgwasi	Vumari	Kizu-ngo	Ishinde	Njoro	Mga-ndu	Mvure-kongei
Agricultural work										
Wife	6.27	6.03	6.87	5.90	6.80	0.94	4.47	6.31	6.17	7.20
Husband	6.50	5.90	6.00	6.00	7.00	0.90	6.00	6.00	5.90	7.20
Child	2.37	2.57	4.72	4.44	0.71	3.19	1.52	3.52	3.97	6.93
Fuelwood										
Wife	2.76	3.93	2.48	2.26	1.66	2.41	5.43	5.77	6.00	6.00
Husband	0.20	0.00	0.91	0.00	0.52	0.00	4.23	0.57	0.00	0.00
Child	2.07	1.63	1.20	2.68	1.26	2.12	2.13	2.16	3.97	4.87
Water										
Wife	0.80	3.09	2.19	2.36	1.06	2.45	5.87	3.84	1.73	2.01
Husband	0.05	0.03	0.85	0.07	0.20	0.89	4.57	0.67	0.00	0.00
Child	0.89	0.71	0.53	1.15	0.55	0.91	0.92	1.12	1.70	2.09
Total										
Wife	9.83	13.05	11.54	10.52	9.52	11.80	15.77	15.92	13.90	15.21
Husband	6.75	5.93	7.76	6.07	7.72	8.79	14.80	7.24	5.90	7.20
Child	5.33	4.91	6.45	8.27	2.52	6.22	4.57	6.80	9.64	13.89

Source: Dimoso, Tables 3.3, 3.5 and 4

As can be seen from the table, in Mvurekongei village, children of school age were working almost 14 hours a day on these three tasks. How can they have any time, or energy, for school? Women in that village were working over 15 hours a day on the three tasks, and presumably more on household tasks. Men were working 7 hours a day, entirely on agriculture and livestock.

In all these villages, women were giving between just under 10 hours a day and nearly 16 hours a day to these tasks – much more than the men, except in one village, Ishinde, where the men were working nearly 15 hours a day.

In the three villages in the most degraded area, the women were also spending up to 6 hours a day collecting fuel wood.

On average in the ten villages, both men and women were spending over 6 hours a day on agricultural work, and children of school age were spending about three hours a day. Dimoso concluded, very plausibly, that the amounts of time they had to spend collecting water and firewood made it very hard to spend much more time in the fields.

In Ishinde village, the women were spending nearly 6 hours collecting water, and the men another nearly 5 hours. It is no wonder they were having difficulty making a success of their agriculture, or dealing with the problems of loss of soil fertility and erosion.

The figures also show that there is wide variation between villages, even when they are close together geographically. Gwanga and Vumari have good sources of water, so its residents do not spend so much time collecting water. In the lowland zone, and especially in Ishinde, the amounts of time spent collecting fuel wood and water are almost unbelievable.

Further reading on open access on the internet

1. *Gender, Growth, and Poverty Reduction: 1998 Status Report on Poverty in sub-Saharan Africa.* World Bank Technical Paper 428, 1999 http://documents.worldbank.org/curated/pt/677841468767650869/pdf/multi-page.pdf

2. *How much of the Labor in African Agriculture is provided by Women?* Amparo Palacios-Lopez, Luc Christiaensen and Talip Kilic. World Bank Policy Research Working Paper 7282, 2015. http://documents.worldbank.org/curated/en/979671468189858347/pdf/WPS7282.pdf

3. *Debunking the myth of female labour in African Agriculture.* Cheryl Doss, IFPRI blog, 2015. http://www.ifpri.org/blog/debunking-myth-female-labor-african-agriculture

4. *Raising the Productivity of Women Farmers in sub-Saharan Africa,* by Katrine Saito. World Bank Discussion Paper 280, Africa Technical Department, 1994 http://documents.worldbank.org/curated/en/812221468741666904/pdf/multi-page.pdf

5. "Women and agricultural productivity: Reframing the Issues", by Cheryl Doss, *Development Policy Review* Vol.36 No.1 pp.1–16, 2017. https://www.ncbi.nlm.nih.gov/pmc/articles/PMC5726380/

6. Lucia Da Corta and Juanita Magongo. *Evolution of gender and poverty dynamics in Tanzania.* London: Chronic Poverty Research Centre, Working Paper 203, 2011. https://www.gov.uk/dfid-research-outputs/evolution-of-gender-and-poverty-dynamics-in-tanzania-cprc-working-paper-203

7. Jeanne Koopman and Iba Mar Faye. *Land Grabbing and Women's Farming, and Women's Rights in Africa.* International Conference on Global Land Grabbing II, Cornell University, 2012. https://www.yumpu.com/en/document/view/30251407/land-grabswomens-farming-and-womens-activism-in-africa

8. Romanus Dimoso. *Environmental Degradation and Intra-Household Welfare: The Case of the Tanzanian Rural South Pare Highlands.* PhD Thesis, Wageningen University, Netherlands, 2009. http://edepot.wur.nl/7443

9. Special issue of *Agriculture for Development* (No. 32, 2017) on *Women in Agriculture: Negotiating and Networking.* Especially the introductory article by Christine Okali and Imogen Bellwood-Howard (pp.2-6), Nozomi Kawarazuka, Catherine Locke and Janet Seeley "Rethinking how Gender Matters for Food Security", (pp.34-7). https://www.taa.org.uk/assets/pubs/Ag4Dev32

Further reading not on open access on the internet

10. Andrea Cornwall, Elizabeth Harrison and Ann Whitehead "Introduction: Repositioning Feminisms in Gender and Development", *IDS Bulletin* Vol.35, No.4 (2004) pp.1–10.

11.   Tibaijuka, Anna. "The Cost of Differential Gender Roles in African Agriculture: A Case Study of Smailholder Banana-Coffee Farms in the Kagera Region, Tanzania". *Journal of Agricultural Economics*, Vol.45, No.1, 1994.

Topics for essays or exam questions

1.   Discuss why women in Africa are expected to work such long hours, in the homestead and in the fields. Is what you describe sustainable in the long term?

2.   Summarise what you know about the division of labour between men and women in growing crops and the care of livestock in one village in Africa known to you. Do the women have separate plots? Are they permitted to keep money from the produce they sell? When the men sell crops, do they share the money with the women? How could these arrangements change in future?

3.   Discuss the problems and challenges in agriculture which face women who are heads of households. Explain why women who are heads of households in Africa are not all poor.

4.   Can you be a successful farmer if you cannot read or write? How does this affect women? What skills should be taught in primary schools that would be useful to women farmers?

5.   The 1998 World Bank "Status Report" [1] reported that men and women worked on separate plots, and that the yields on the plots cultivated by women were higher than on those cultivated by men. From this they concluded that there would be big increases in production if more women adopted the same techniques as men. Discuss what is NOT taken into account in this advice.

6.   Discuss the possible experiences of groups of women who start agricultural or livestock enterprises. How might a formal organisation in a co-operative assist them?

7.   How can women get access to credit for agricultural or livestock enterprises?

8.   Discuss the view that the position of women in all aspects of life, including agriculture, is more influenced by films, or TV or radio "soap operas" than by government policies.

9.   What innovations or changes would make the most difference to women's lives in rural areas in Africa?

10.   Explain how, when a husband dies, or a marriage breaks up, women often have difficulty keeping the land they have been farming. What changes should be made to ensure that women in such circumstances have access to land?

CHAPTER 12

Agricultural Policies

TWELVE ESSENTIALS FOR SUSTAINABLE AGRICULTURE

This final chapter brings together some of the key themes discussed in earlier chapters, and summarises the policies that are needed to promote a more productive, and sustainable, agriculture. It is addressed to those who are able to influence policies that relate to agriculture in Africa – in governments, donor agencies, NGOs, farmer organisations and businesses which are prepared to invest in agriculture for the long term.

Attentive readers may already have noticed a difference in emphasis between the previous two chapters. Chapter 10, on green revolutions and genetically modified seeds, showed how small-scale farmers may not need chemical inputs to be successful. It did not claim that they would always succeed, or that they could do so without extra labour input, or that their yields would be as high as when fertilisers and other purchased inputs are used. But it did suggest that many can make a living that way, and it emphasised the importance and value of many of the technologies that are built into traditional agricultural systems – the careful selection of seeds from the best plants for use the following year, the need to plant drought-resistant varieties, the value of intercropping as a means of lessening soil erosion and reducing the risk of pest infestations.

Chapter 11, on gender, showed that women in traditional families are disadvantaged in many ways. This has severe economic costs, because the techniques most appropriate to women's needs are not always used. There are social costs too, when girls do not have time to go to school and women work very long hours. So when it comes to gender, governments and other agencies are often working to challenge and to change traditional cultures, and to give more recognition to the contributions that women make. This may involve encouraging women (and men) to use their democratic powers at national level, and in smaller organisations at local level, to do what they can to get better treatment. Sometimes this will involve working with other women in the local situation, for example to grow a new crop or to set up a small maize or rice mill.

So the position of this book is that there is much of value in many traditional practices, and the relevant skills and knowledge should not be lost. But there are also traditional customs and practices which discriminate against women or other groups, or which are inefficient, and these need to be challenged. In addition there are technical innovations of many sorts which, if carefully applied, can result in greatly increased production and improved standards of living of the rural population as a whole.

The case study in Chapter 11 presents information from the PhD thesis by Romanus Dimoso, about an area of Tanzania, the South Pare Mountains. This area used to be highly productive, but now the best soils have lost their nutrients or been washed away, forests have been cut down, and water supplies and fuel wood have become so scarce that some women and children are walking for six hours a day to collect fuelwood and more than two hours a day to collect water. Erosion is a major problem. Farmers are being forced to leave the area because otherwise they cannot survive – a problem made worse by continuing population growth and by climate change.

Something similar happened in the Ismani area, North of Iringa, in the 1970s. This was hailed as the new frontier for mechanised agriculture, comparable to the plains of the Mid-West of the United States in the nineteenth century. Deep ploughing by tractors led to the loss of the thin layer of good topsoil. This was followed by erosion, and now the area is a wasteland which will take years to recover, if it recovers at all. There are very real dangers of similar erosion taking place if marginal areas are opened up and ploughed with tractors, without very careful (and sometimes expensive) measures being implemented to protect the land.

This chapter, and the book, concludes by emphasising twelve points made in earlier chapters, and converting them into points for action. They are:

1. Farmers must have security of tenure over the land they farm. Without this they will have little incentive to invest in improving their land, or even in maintaining its fertility. It is therefore crucial to long-term sustainability.

2. Farmers need to grow crops which are high yielding, drought tolerant, disease resistant, of high value, and make maximum use of farmers' expertise and family labour.

3. Livestock are as important as crops – more important in some situations. They should get much more attention than they do at present.

4. Technologies need to be chosen which utilise farmers' resources to the full, and avoid exposing them to excessive risk. Wherever possible, the innovations, and the ways in which they are implemented, will work best if they are developed in close association with those who will use and benefit from them.

5. More than half of those who work in agriculture in Africa are women, and they do more than half the manual work. Those who have influence over policies need to work with women, at least as much as with men.

6. Investments outside agriculture may in some circumstances have greater impacts than investments directly in agricultural production. For example, investments in feeder roads may reduce the costs of marketing; investments in water supplies for domestic use may reduce the huge amounts of time at present spent on collecting water, and largely eliminate the dangers of poor health arising from the use of impure water.

7. Where markets work well they are cheap and efficient. But they do not always work well, so there are situations where the state needs to be prepared to intervene. Whether services are provided by the private sector or by the state, they need to be underwritten by efficient and honest systems of inspection and regulation. Overall, if the marketing system for a crop is not effective, farmers are unlikely to increase production of that crop.

8. Credit is an important component of agriculture. But for small farmers it can be expensive and risky. In some situations it may be possible to provide credit to groupings of small famers. Otherwise it may be preferable for a government to make direct subsidies.

9. Most innovations have not depended on formal research. But there are huge potential benefits from such research, if it is done well. The state and donors need to strengthen the research services and recognise that they can make a huge difference to agricultural production.

10. The extension service is an important resource, but it needs to be pro-active rather than re-active, and where possible to work with groups of farmers engaged in specific proposals for improvement, rather than with individual farmers.

11. To overcome diseconomies of scale, small farmers should be encouraged and assisted to work together in groups, for accessing credit, inputs and markets as well as facilitating extension and other services. The groups may take the form of co-operatives, or of contract farmers supplying raw materials to agricultural processing companies.

12. The government's agricultural policies should be closely linked to a national strategy for improving public health and nutrition, which should be the end goal of a productive agricultural sector.

These points are explained in more detail below.

1.   Farmers must have security of tenure over the land they farm.

Land has featured in every chapter of this book. The first chapter starts with a discussion of soils, and how crops grow in them. It shows how soils can easily be eroded, or lose their fertility, and ends with a note about the likely impacts of global warming. The next chapter, on the uses of land, introduces the concept of conservation agriculture, protecting the soil by avoiding unnecessary deep ploughing, which buries nutrients needed by plants deep in the soil, and leaves the soil exposed and vulnerable to wind and rain.

Farmers on any scale, whether large or small, need security of tenure – so that they know that their land will not be taken away from them provided they use it productively. This is of obvious importance to those farming on a medium or large scale, which is why the main discussion of land tenure is in Chapter 6 (on larger farms). But security of tenure is equally important for small farmers. They will have little incentive to invest, to preserve and enhance the fertility of the land they farm, and to protect it from erosion, unless they are sure that they will be able to farm it for the foreseeable future.

If small farmers do not have secure tenure, and are thrown off their land, not only will their contribution to national agricultural production be lost, but they and their families will be at great risk of becoming destitute. Either they will join the ranks of the rural poor and landless, or they may be obliged to migrate to the towns, where they are likely to add to the numbers of urban unemployed, putting social cohesion and stability at even greater risk.

Secure forms of tenure may be freehold, where the land has been surveyed and mapped, and the farmers have titles or certificates which confirm their rights to it; or it may be leasehold, where a landowner, or an organisation such as the national government or a village government is confirmed as the owner of the land and guarantees individual farmers the rights to use it.

2. **Farmers need to grow crops which are high yielding, drought tolerant, disease resistant, of high value, and make maximum use of the farmers' expertise and family labour.**

With heavy pressure on cultivable land resulting from the high rate of population growth, farmers need to maximise their return per hectare and per input of labour. This means choosing enterprises which can be intensively farmed, and using production systems which make the fullest possible use of family labour and expertise.

There are notes on the main crops in Chapter 2, and important discussions of the motivations of small farmers in Chapter 5 and of larger farmers in Chapter 6.

Farming, small or large, is a business. Farmers take decisions which involve different levels of risk, and they live with the consequences. For small farmers, food security is the first essential – they must be able to access sufficient food to get through long dry seasons without having to buy food when purchase prices are at their highest, just before the next harvest. To do this they will lessen their risks: by planting many different crops, often several together in the same field, by planting at different times, by planting different varieties (some more drought resistant than others), and by keeping livestock which are adapted to survival in harsh circumstances.

But productivity also matters, especially when good land is scarce. So farmers need to be able to access seeds and planting materials that will give them good yields, and they need to know how best to get high yields – either through appropriate use of purchased fertilisers and other inputs, or through organic methods, including the use of green manure, plants that repel insects and other techniques discussed in Chapter 10. When land is used intensively for many years, farmers need to ensure that organic matter is returned to the soil so that fertility is not depleted.

Many indigenous crop varieties have good properties. But so do many hybrid seeds and so, potentially, do some genetically modified crops. The resulting cooked food may not taste so good, but the crops may grow quickly, give good yields, store well and resist some of the most devastating diseases. There are places for both traditional and modern varieties on a successful farm.

3. **Livestock are as important as crops – more important in some situations. They should get much more attention.**

Cattle, goats, sheep, rabbits, chickens, ducks, geese, turkeys, fish, and bees make major contributions to the economies of most African countries, as do animals that live in the wild. Many animals and birds can survive periods of drought.

In Tanzania livestock farming contributes nearly two-thirds as much to GDP as crop agriculture. Often livestock and crops are integrated, in "mixed farming" systems. In very dry areas, where the risk of crops failing completely is severe, animal husbandry is often the optimal use of land.

Animals are sources of protein and make varied diets possible. The meat and hides they supply are important contributions to the economy. They pull carts and ploughs. They improve the soil and the crops that grow on it. They are an insurance policy and a bank for their owners.

Nomadic pastoralism has the advantage of using available grass and other forage, especially in areas that are unsuited for annual cropping, but it has the disadvantage that land may be overgrazed, and that conflicts may develop with growers of crops, even though the animals contribute manure which improves the fertility of the land. Pastoralists have frequently been misunderstood, ignored, or even discriminated against, often on the basis of allegations that have no foundation in reality, such as the claim that they only sell cattle when they are old. The reality is that pastoralists are well aware of the values of different ages of cattle: they sell them to maximise their incomes, and they are an integral part of the rural economy.

Ranching, where land is enclosed by fences, so that the pastures can be used in rotation to prevent over-grazing, has proved to be difficult to manage profitably in many parts of Africa. However, nomadic pastoralism is threatened by rapid increases in areas of land allocated to crop agriculture, and by losses of water sources close to pastures, especially in the dry season.

Controlled grazing, and intensive production of cattle, sheep, goats and small animals and birds, should wherever possible be actively encouraged and supported as part of the overall farming system.

4. Technologies need to be chosen which utilise farmers' resources to the full, and avoid exposing them to excessive risks.

The basis of the technologies relevant to farming was summarised in Chapter 2: minimal disturbance to the soil before crops are planted, intercropping so that a fast-growing crop provides cover and reduces the risk of wind and water removing the topsoil, crop rotations including the cultivation of fodder crops or other crops that fix nitrogen from the air. Also the cultivation of a large number of crops to minimise the risk of total crop failure, the use of green manure, the creation of ridges or terraces along contours, and of ditches to retain water, better forms of storage, and use of the most appropriate equipment or machinery.

Technologies which optimise the use of farmers' resources are relevant and necessary on any size of farm. Heavy machinery which replaces family labour without necessarily raising yields may not be the most appropriate choice. Investments in agriculture should always be planned to be sustainable. In most cases that will be because they are profitable and generate income which enables them to survive for the long term. For this to happen, the technologies must be appropriate and support systems must be in place – supplies of inputs and spare parts, technical advice, efficient markets and market information, and short and long term weather forecasts. Conservation technologies and soil conservation strategies should be built into the production system from the start of every project, and then maintained.

Government and donors should examine carefully any aspects of their interventions which, while benefiting some, may make things worse for others.

For example:

a) advising farmers to switch from multiple cropping in favour of planting a single crop, as described by Dawson and others for parts of Rwanda (see the discussion in Chapter 10), may produce a lot of maize in a good year, but it will increase the risk of total crop failure in a bad year, and it will leave many people without work for much of the year, and may drive them into destitution.

b) large-scale irrigation schemes (as described in Chapter 3) may result in a good water supply to farmers near the start of the water distribution system, while depriving water-users lower down the system of the supplies they need. This can lead to very unpleasant disputes. A smaller project, planned in consultation with all the farmers who are affected by it, may well be more productive in terms of overall yields and less likely to lead to disputes than large-scale centrally managed irrigation projects.

5. **More than half of those who work in agriculture in Africa are women, and they do more than half the manual work. Those who have influence over policies need to work with women, at least as much as with men.**

Chapter 11 shows how much of the discussion about women and their contributions to agriculture in Africa has been dominated by a series of myths and half-truths. In contrast, the most recent evidence suggests that women work about the same numbers of hours in the fields as men, and, increasingly work together with men – the concept of separate "women's fields" and "men's fields" is breaking down. Moreover, women heads of households are often successful farmers – which is hardly surprising when many of the other tasks they undertake require similar forethought and planning for the future.

There are policies that can build on this potential, such as making it easier for girls to attend schools, ensuring that women are given fair opportunities for promotion in companies and government organisations, and employing more female extension workers. But the change that can make the most difference is simply to recognise that women can make contributions as successfully as men, and to ensure that in every programme they have opportunities to do so.

6. **Investments outside agriculture may in some circumstances have greater impacts than investments directly within agricultural production.**

Here are some examples of investments outside agriculture:

a)   Investment in domestic water supplies may release large amounts of time previously used for fetching water, which can be applied to intensifying agricultural production. Dimoso's data at the end of the previous chapter showed huge amounts of time spent collecting water and firewood. This is, perhaps, an extreme case, but his conclusion that too much time is spent on this will still be valid even if the time saving is less than it would be in those particular villages. If a water supply is installed, the infrastructure must of course be maintained. Africa is littered with disused water towers, broken pipelines, and taps which do not supply water, because nobody had the skills to repair a system when it broke down, or the spare part that was needed, or the money to buy fuel for a pump – or simply because no one reported the problem to the relevant authority. Domestic water supply projects, just like any other projects, must be sustainable. This means either that money for maintenance and repair must come from central government, or from local government, or, perhaps most reliably, from some system of taxation on those who benefit: the farmers, their family members, and other residents who will no longer have to walk for hours to collect water. Whichever way the money is raised, this must be planned, and anticipated, from the outset.

b) Alternative fuels. The time-saving argument here is similar. The fuel may be electricity (though it would require a large array of solar panels to supply a whole village), or fuel wood plots or planted forests (as, for example, in the Njombe area of Tanzania, or in places where large quantities of firewood are used for curing tobacco). More efficient cooking stoves use less fuel, and also generate less smoke, so there is a double benefit. Gas stoves are already used in urban areas, and would be efficient in many rural areas (in Africa, gas is likely to become increasingly available). Perhaps, most important of all, steps should be taken to limit the production of charcoal for urban areas, because, as Dimoso points out, charcoal producers destroy whole trees, whereas those who collect wood for domestic use usually just cut branches – and the trees survive.

c) Better roads and forms of local transport will save time and energy, and can open up new markets. This includes better feeder roads, tracks and bridges, and also carts pulled by oxen, or two-wheel tractors.

d) Better storage systems, which can be sealed and fumigated, will ensure that crops that are dry when stored can survive for several months, and be used when required for consumption or sold when prices are high.

7. Where markets work well they are cheap and efficient. But they do not always work well. So there are situations where the state needs to be prepared to intervene. But whether services are provided by the private sector or by the state, they need to be underwritten by efficient and honest systems of inspection and regulation. Overall, if the marketing system for a crop is not efficient, farmers are unlikely to increase production of that crop.

In Chapter 4, on innovation, it was noted that many innovations or new products spread rapidly on their own, without direct state intervention. Mobile phones are an obvious example. Another is the spread of round potato production in the Mbeya and Njombe areas of Tanzania, discussed in Chapter 9. Another example is the spread of ox ploughs and carts in many parts of Africa.

Markets can also set the wrong signals, as noted in Chapter 7. If a farmer is not paid an agreed price for a crop, or receives very little, it is to be expected that he or she will be reluctant to grow that crop in the future. This may occur because a private trader, or a co-operative or a marketing board, is exploiting the farmer, or because of inefficiency (e.g. if a buying agent runs out of cash), or because for some reason there is no market for that crop (for example a surplus of a perishable crop such as tomatoes).

Markets work best when traders have long term relationships with those they sell to, and have access to sufficient working capital to make their purchases. In the years of structural adjustment in countries such as Tanzania, the marketing boards and their successor bodies were disbanded. The private sector was expected to take over. But many traders did not have the skills, or the contacts, to appear credible to overseas buyers. This was one of the causes of the agricultural crises of the 1990s.

The state can assist. For example, a modest tariff on imports of rice can encourage local production. Nigeria became the largest producer of cassava in Africa after the state intervened: one of its actions was to force millers of wheat flour (almost all its wheat is imported) to include 10% cassava flour.

Marketing, and hence agricultural production more generally, often depends on efficient and honest systems of inspection and regulation. Farmers need to know that weighing scales are accurate and that their crops are being correctly graded. Inspection is needed to ensure that crops meet quality standards, especially for exports. Thus cotton ginneries need to be inspected to ensure that the cotton lint they produce is not contaminated with soil or other impurities. Processed foods must be inspected to ensure that their contents are as described on the packaging, that they are not contaminated, or dangerous in any way, or a risk to human health (for example groundnuts that are not properly dried can contain aflatoxin, a dangerous poison). Animal products pose many of the greatest risks to health.

But inspection is costly, and open to corruption. There are never sufficient inspectors. So where it is possible, it is better to strengthen market forces, for example by giving farmers information about prices in different markets. These days this can be provided by use of mobile phones. Farmers need to know where they can get higher prices if they sell crops of higher quality. Or to know which traders cannot be trusted.

8. **Credit is an important component of agriculture, but for small farmers it can be expensive and risky. In some situations it may be possible to provide credit to groupings of small famers. Otherwise it may be preferable to make direct subsidies.**

Credit was discussed at the start of Chapter 8. For large farms, where a loan may be secured against the assets of the farm, the issues are straightforward – though those involved in large-scale farming may not wish to risk losing their assets if something goes wrong and they are not able to repay a loan.

But for small farms, the overhead costs are very high if loans are given to each farmer individually, so interest rates are high and the farmers feel they are being exploited. Banks may be reluctant to lend to farmers, and farmers may be reluctant to take on the responsibilities of loans. It is for such reasons that few agricultural banks in Africa have been successful.

Where lending to small farmers has succeeded, the loans have often been given to groups of farmers, or to co-operatives and not to individuals. Then if an individual does not repay a loan for any reason, the other farmers in the group are still responsible for the repayment. So they have a very strong incentive to ensure that every farmer complies.

Because of the difficulty of establishing an effective system of credit for agriculture, with high loan-recovery rates, governments have often resorted to grants, for example for fertiliser. The bureaucracy is much less. The fertiliser is provided. And most of it will be used, though not always where the government expected it to be used.

9. There are huge opportunities for innovations based on
    research. The state and donors need to strengthen the
    research services and recognise that they can make a huge
    difference to agricultural production.

Agriculture is probably the only area of the economy where the colonial
powers handed over to newly independent states an infrastructure of
research institutes and scientists trained in the various specialist areas
at the cutting edge of technology, on a world basis. It stayed that way
in Tanzania until the 1980s, and then suffered after the budget cuts
associated with structural adjustment, when salaries did not keep up
with inflation.

Even so, the basic research infrastructure survives in many places.
African countries still have the wherewithal to develop new varieties
of seeds or planting materials, and to test them. They have the capacity
to test new methods of animal husbandry, and of controlling animal
diseases. They can develop different methods for controlling plant and
animal pests and diseases, including through technologies based on
biological control. They employ agricultural economists, sociologists
and agricultural engineers, who can work with groups of farmers
to understand traditional technologies, and consider how best to
improve them.

But across the Continent, the research service is only just surviving,
and not everywhere. For agricultural research is labour intensive,
long term, and involves teams of experts who coordinate their work.
It also requires constant attention to detail, for example to ensure that
samples are labelled and stored safely, that activities are carried out as
planned when the research was devised, that records are kept up to
date, that conclusions are written up and disseminated, and tried out
on farmers' farms in a wide variety of locations to ensure that there are
no unexpected outcomes. If there are vacancies in key teams, or delays
in replacing scientists who leave, if there are equipment breakdowns, or
failures to get key activities completed on time, if the quality of record
keeping slips, or if research is never completed, or is not tried out on
farmers' farms, then much of the benefit that should flow from the work
will be lost.

There are new players to deal with too. From the 1970s, international
organisations started supporting certain kinds of agricultural
research, such as AGRA (Alliance for a Green Revolution in Africa,
supported by the Bill and Melinda Gates Foundation), or the many
research organisations that comprise the CGIAR group (originally the

Consultative Group on International Agricultural Research). Private companies are also involved, because many of these organisations have links with the green revolutions which were successful in Central and South America and in Asia, where there were commercial profit possibilities for companies that sold seeds, fertilisers, insecticides and weedkillers. They would like to see the same technologies rolled out in Africa.

The research service should be the jewel in the crown of Ministries of Agriculture in Africa, because the capacity to develop world class technologies is there, and the rewards are potentially huge. But this will happen only if research institutes and projects are funded for the long term, if the quality of the research itself is maintained, and if resources are available for the researchers to travel in the field locally, and to visit other scientists working on similar projects at conferences and workshops. It is time, in many African countries, for a hard-headed review of the situation, for decisions about priorities, and then for campaigns to get the resources and staff to make the work happen.

10.   **The extension service is an important resource, but it needs to be pro-active rather than re-active, and to work with groups of farmers engaged in specific actions for improvement, rather than with individual farmers.**

Chapter 9 was harsh on the present organisation of extension services in most parts of Africa – because it is expensive and labour intensive, and often it is unclear what it achieves.

The biggest criticisms are of "generalist extension workers" based in villages who are left, or more or less left, on their own, to advise farmers and help them respond to problems as they arise.

Cost-effective extension needs to be based on specific campaigns, addressed to groups of farmers. The campaigns may be related to specific crops, or they may be generic - for example on making better use of water, conservation agriculture, organic farming, new methods of marketing, or integrating crop and animal husbandry.

Training should target groups of farmers, not individuals. It can take place in farmer training centres, at schools or in village centres. The training should start from the issues voiced by the farmers, and should promote specific technologies only if these are seen as the most appropriate means of responding to the problems.

Above all, those running extension services need to understand the conclusions drawn from Chapter 9. Research should start from discussions with extension workers or groups of famers about the problems they face. It should recognise the traditional agricultural practices that have developed over time to protect the soil, minimise the risks of failure and starvation, and use the labour available in the best ways. And then it should improve them. *Work with farmers should be at the centre of all agricultural policies.* Those who take technologies and products from outside and try to direct farmers to use them, without understanding how they may relate to the technologies that are already in use, are very likely to do more harm than good.

11.   To overcome diseconomies of scale, small farmers should be encouraged and assisted to work together in groups, for accessing credit, inputs and markets as well as facilitating extension and other services. The groups may take the form of co-operatives, or of contract farmers supplying raw materials to agricultural processing companies.

Very small farmers working on their own are unable to benefit from economies of scale. They can overcome this constraint by working together in groups for specific purposes. The old multi-purpose co-operatives have a bad reputation in many parts of Africa: they tried to take on too many different tasks, lost their personal touch with individual members, and became known for mismanagement and corruption.

However, smaller groups of producers who know and trust each other, uniting for specific purposes be it supply of credit or inputs, use of equipment, access to markets and services including extension, storage of crops and value addition by processing, have been much more successful.

A form of producer group which has shown great potential is where a company contracts small farmers to supply raw materials for processing and marketing, as outlined in Chapter 8. The company provides the farmers with inputs and technical support, and offers a guaranteed market and a confirmed price for their production. The farmers use their land and labour to supply the crop that the company needs: they form a group for communicating and negotiating with the company, and ensuring a fair deal from which both sides benefit.

Governments, aid agencies, NGOs and investors should encourage this approach to group formation, offering training and support to ensure that the contracts are fair and transparent, and that both buyers and sellers adhere faithfully to their part of the agreement.

12. The government's agricultural policies should be closely linked to a national strategy for improving public health and nutrition, which should be the end goal of a productive agricultural sector.

Two crops heavily promoted in Africa are perceived as killers in Europe and America, where governments are trying to reduce their consumption or eliminate it entirely. Cigarette smoking is a cause of cancer, not just for those who smoke but for those who live in the same houses or workplaces and breathe the smoke. Smoking is also associated with heart disease and with strokes. A high dependence on tobacco farming makes it hard for governments to discourage smoking.

Sugar is the darling of the food-processing industry, because it is cheap and is a reliable preservative which stops foods deteriorating, and also because, like tobacco, it is addictive. Once you get used to having sugar in your diet, it is very hard to stop. Yet consumption of refined sugar is the main cause of obesity, which causes heart diseases, diabetes, and many other conditions, as well as lessening physical strength and mobility. It is probably not possible to eliminate sugar entirely from most diets, but countries in Europe are seriously considering imposing taxes on bottled drinks such as colas with high levels of sugar. Some have already done so.

There are other public health concerns. Charcoal and wood burning are also causes of cancer, as are the fumes from petrol and, especially, diesel engines.

Governments should promote healthy eating – at least one piece of fruit per day, a balanced diet with protein and minerals from meats or legumes, vitamins from unprocessed vegetables, and carbohydrates from cheap, reliable sources, such as cassava, sweet potato or banana.

The ultimate target of everything in this book is a well-fed and healthy population in both rural and urban populations, on a sustainable basis. The steps outlined above, and set out in detail earlier in the book, can improve the productivity and sustainability of agriculture in Africa. They can provide most of the food needed in the rapidly growing urban centres, as well as raw materials for industrial processes. Above all, they can provide the purchasing power for a wide range of consumer products, and products linked to the construction industry, transport and local small-scale crafts and workshops, and so make possible a form of industrialisation in Africa in which everyone benefits.

END GAME
A Short Quiz

What are your understandings of family-based agricultural units?

To end this book, here is a short quiz by Frédéric Kilcher, designed to bring out the strengths and the opportunities offered by small farmers in Africa. It is followed by his short guide to how this can be done.

Answer the following questions (True / False) to test your vision of farmers:

Statements

1.	Farmers are not well educated so it is OK to tell them what to do.	True/False
2.	Modern inputs lead to better yields so farmers must accept them.	True/False
3.	Farmers' practices are outdated so they should replace them with new practices.	True/False
4.	Farmers have sufficient experience to handle their crops, livestock activities and to manage their households.	True/False
5.	Farmers learn first from their fellow farmers before they learn from extension agents, researchers, agro-dealers, etc.	True/False

6.	It is irrelevant to study farmers' practices and performances. These are not good so they should be replaced with what scientists identified as better practices/performances.	True/False
7.	Labour calendars allow identifying the level of utilisation of the family labour each month, hence the farmers' ability to integrate new crops, new practices, etc.	True/False
8.	Cash-flow charts cannot be consolidated for smallholder farmers who do not keep records.	True/False
9.	Improved yields always translate into improved livelihoods.	True/False
10.	Improved yields can cause losses of incomes to farmers.	True/False
11.	Cash-flows allow us to highlight periods when farmers suffer financial stress.	True/False
12.	Cash-flows allow us to identify farmers' capacity to invest in new practices/inputs.	True/False
13.	Improved productivity allows sales to be increased, so it always results in increased net incomes for farmers.	True/False
14.	It is possible to assess farmers' net income.	True/False
15.	Projects and governments understand the farmers' financial and labour constraints, and know what farmers can take on board (new practices, new tasks, new expenses).	True/False
16.	Farmers are keen to improve their financial management.	True/False
17.	There are places where smallholder / subsistence farmers have adopted better financial management, and this has had a positive impact on impact on their livelihoods.	True/False
18.	There is a database of benchmarks on African farmers' performances.	True/False
19.	All smallholder farmers are more or less the same and need the same support activities.	True/False
20.	Agriculture has the potential to drive the economic growth of African countries.	True/False
21.	Farmers are lazy and stubborn. It is better just to tell them what to do.	True/False
22.	Increased sales are a good indicator of a farmer's improved livelihood.	True/False
23.	There is enough land for all farmers to develop their activities.	True/False

24.	Farmers can expand their production substantially if they get enough inputs.	True/False
25.	Using certified seeds always leads to improved yields.	True/False
26.	Farmers do not want to progress.	True/False
27.	Smallholder farmers cannot thrive under the current circumstances (practices, education, etc.).	True/False
28.	Farmers have access to well-adapted packages of inputs / services / information sufficient for them to progress.	True/False
29.	If the following support activities (such as inputs, mechanization, finance, information, marketing, infrastructure) are in place, farmers will develop. It is not necessary to combine and coordinate these interventions.	True/False
30.	Farmers are not able to identify the best ways forward for their own human and economic development.	True/False

Answers:

Facts: True or False

1.	**False.** They may not have high levels of education but they are the ones who have to deal with the consequences of their choices/actions, so they have a right to decide what they want or do not want to do. A low level of education does not mean that they have no experience or intelligence. This must be respected and accepted before any interaction with farmers occurs.
2.	**False.** Modern inputs do not always lead to better yields (when not adapted to the specific conditions on a farm, not supplied on time, not of good quality, or not well combined with practices or other inputs). Farmers must understand the likeliness of these inputs generating the declared result so that they can decide on the risk involved. It is therefore not right to impose inputs on farmers. On the contrary, farmers should be taught how to assess the viability of using inputs (financial calculations, in-situ testing, etc.).
3.	**False.** Practices may be outdated when compared to advice from research stations, but this does not mean that these practices are not adapted to a farmer's context, abilities and situation. New practices may not fit into the farmers' systems (cost, labour intensity, etc.). Any replacement must be based on farmers' understanding of the feasibility and impact of using new practices. Again, farmers are the ones to cope with the embedded risk of changing. Imposing changes without guaranteeing compensation if the new practice fails is not acceptable.

4.	**False.**
	Farmers have being doing this for hundreds of years. Their education may not be optimum but their experience and strategies allow at least stability, and sometimes progress. However, it is true that the circumstances are changing at an increasing speed (pressure on land, climate change, increased needs, etc.) and past strategies/ practices may not be suitable in the future. Helping them to adapt is a reasonable thing to do.
5.	**True.**
	Studies show that farmers learn most from their fellow farmers, a lot from agro-dealers, and only a little from extension agents. This shows that they are able to change and that the service provided by extension is seldom a useful source of information.
6.	**False.**
	Without an understanding of the prevailing practices and performances it is not possible to identify new practices suited/ feasible for farmers, nor to determine whether these "new" practices will lead to the performances expected.
7.	**True.**
	Farmers depend primarily on their own labour. Every month, each of the family members can provide a specific quantity of labour. Each task requires a specific work input. It is possible to measure the labour availability and the work-input needed. The measuring unit is the number of "Person-Days". Building a labour calendar makes it possible to identify the farmers' labour balance (shortage or excess of labour in each month). This will allow an understanding of whether the farmers can take on board new crops or practices. If they are short of family labour, they will have to pay for hired labour (if their cash-flow allows) or limit the care given to crops/livestock activities/family needs. So labour calendars are vital to assess the farmers' ability to progress (and the impact of mechanisation).
8.	**False.**
	Farmers may not have records readily available but under some circumstances and with a careful approach, it is possible to re-create farmers' cash-flows. Their accuracy may not be 100% but even if only 60% accurate, they help to highlight the families' financial dynamics, the need to build up savings, and their net incomes.

9.	**False.** Farmers' livelihoods depend on their cash and food availability and also on the level of stress on the family labour force. Improving yields usually will require spending more money (for inputs, services or casual labour) and using more labour. The output of a crop/livestock activity is never 100% guaranteed. The increased production will not result in increased incomes if expenses eat away much of the additional income, or if the produce cannot be sold in good condition for a good price.
10.	**True.** To generate increased yields, farmers may be pressed to use large amounts of money, to buy inputs, labour, services (especially post-harvest). Farmers may not be able to store the produce in good conditions and may therefore lose a substantial proportion before sale. Having large volumes for sale without an adequate marketing strategy can lead to losses. For instance, tomato yields in Morogoro, Tanzania, were extremely high in the 2016-17 season but the market could not absorb all that production so prices dropped and many farmers faced severe losses.
11.	**True.** The cash-flows show the monthly incomes (from different crops, livestock activities or other income generating activities) and expenses (for the different income generating activities and for the family needs) of a given household. They therefore identify the months for which this household is facing expenses higher than its income. It also can show the situation of that household if positive balances are saved to cover the cash needs of months with negative balances. So cash-flows demonstrate: (1) the household's capacity/incapacity to manage its financial needs over a year and (2) the cumulative surplus or deficit that this household is building up over a year.
12.	**True.** If the cash-flow shows cash left over at the end of the year after having paid for all direct production costs and basic family needs, it provides evidence of the family's capacity to invest in new practices or to purchase inputs or other needs.
13.	**False.** Improved productivity does not always translate into increased sales (e.g. there may be post harvest losses). Increased production might have been achieved with high spending on inputs, labour, services, etc. So net incomes are not necessarily higher.
14.	**True.** Cash-flow analysis demonstrates the cash that a family might save after having paid for all direct costs and family needs. The remaining money is the net income that the family can use for productive investments (in equipment, land, etc.) or for family needs (housing, leisure, etc.).

15.	**False.**
	Most governments carry out population censuses and agricultural surveys. However, these surveys are often limited to long quantitative questionnaires. They do not allow much opportunity to build up an understanding, or benchmarks, about farming systems. This limits the possibilities of designing adequate policies and engagement strategies.
16.	**True.**
	When talking with farmers, it is obvious that their wish is to improve their and their children's lives. They all are frustrated by living on a razor's edge year after year and see limited improvement in their existence.
17.	**True.**
	Assessments of AFD-funded projects (French Development Agency) in Burkina Faso, Benin, Niger, Mali show that it has been possible to persuade farmers to adopt financial management tools, with impressive impacts on their food and financial security, and their decision-making and investment capacity.
18.	**False.**
	There is no such database in Tanzania. The only African country with such a database is South Africa.
19.	**False.**
	This is a dangerous statement that could lead to uniform interventions un-related to the realities of individual families. Farmers do not all have the same levels of resources (land, labour, money, skills and experiences). They are not all at the same level and do not all share the same objectives. In any village, there are: (1) the farmers who struggle to cover their basic needs, (2) the farmers who can invest in agriculture (equipment) or business, (3) the ones in between.
20.	**True.**
	80% of the agriculture in Africa is undertaken by smallholder farmers. They are able to start producing a crop with limited initial resources. They are also much more resilient than farms depending essentially on external capital. Limited improvements in smallholder farmers' systems can generate very significant impacts (yields, profitability, investment capacity, etc.). African smallholder farmers have the potential to produce much more, provided they are given the right opportunities and working environment (markets, training, inputs, policies).

21.	**False.** Considering the harsh and changing environment, farmers would have disappeared if they were lazy (unable to stretch their efforts to get over problems) or stubborn (unwilling to adapt their ways when their environment requires changes). Farmers may not accept what extension workers tell them because they are aware of risks involved or do not see the benefits, nor believe that a sufficient part of the constraints which would prevent them to succeed have been lifted.
22.	**False.** Increased sales will only translate into increased incomes if the prices are good. In addition, increased incomes may be offset by increased expenses (production costs, post-harvest costs, marketing costs, etc.) or more labour. So increased sales are not a reliable indicator of improved livelihoods.
23.	**False.** There is much arable land in many African countries. However, only some of that land is suitable for economically viable agricultural production (considering the quality of the soil, reliability of the rainfall, and distance from access roads). Furthermore, the level of population growth implies that soon, new generations of farmers will need new farms and larger farms. Land availability is limited but land needs are growing exponentially.
24.	**False.** Inputs are not the only production factor on which increased production depends. Labour, services, marketing, infrastructures, policies, etc. are also required. In addition, farmers' access to land and other resources is not unlimited or costless. So increased production requires much more than inputs.
25.	**False.** Improved seeds require a combination of production conditions to deliver the expected yields: soil fertility, timely planting, irrigation, weeding, etc. Often, farmers are not aware of the combination of practices needed, or do not have the resources to manage the crop.
26.	**False.** Farmers want to improve their lives. They are fully aware of the changes in their environment (climate, policies, consumption patterns, etc.).

27.	**True.**
	Farmers have much more knowledge and skills than one might believe when considering the outputs of the subsistence strategies in which most are engaged. They are able to adapt to many changes and challenges. But when farmers are exposed to challenges and changes that go beyond their capacity to adapt, they cannot cope. This is the situation to which African farmers are increasingly exposed. It implies that, for farmers to be able to thrive, they need better access to knowledge and skills... compatible with their pre-existing knowledge and skills.
28.	**False.**
	In many African countries, many "public goods" such as supplies of inputs and services, training, information, do not reach all farmers.
29.	**False.**
	It is clear that if production is developed without simultaneous development of the marketing channels, the prices will collapse and producers will not continue their efforts. Production and marketing are part of a system (the value chain) and must be tuned with each other. When the different components are not well coordinated the outputs cannot be optimum. Agricultural development requires integrated well-coordinated sets of interventions.
30.	**Partly true.**
	Farmers have adapted – without external help – to many changes (new crops, evolution of the policies, changing environment, etc.). This capacity to adapt shows that they are able to analyse their situations and identify solutions to their constraints. However, when changes happen too fast, their capacity to adapt may not be sufficient. But with additional training they can learn how to play a strategic role in identifying solutions to their challenges.

How did you get on?

30 good answers: You know smallholder farmers as if you were one yourself!

20–29 good answers: You know smallholder farmers fairly well but be careful not to underestimate them!

10–19 good answers: You have a good basis for understanding small farmers but you underestimate them and may not realise what they can achieve.

0–9 good answers: Do you really believe that farmers are lazy, stubborn and under-educated? Are you sure you want to work in rural development?

Tips for working with farmers

- Trust their desire to improve their lives.
- Believe in their capacity to learn.
- Acknowledge their skills and knowledge, to produce, assess risks, manage resources, test solutions, handle complex situations, etc.
- Learn how to add to their existing knowledge (adult learning approach) and not to ignore it.
- Address them on an equal basis without putting forward your academic knowledge.
- Value their rationale and their possible contribution to the identification of solutions.

Models of, or guides to, how farmers make decisions

From this it is a small step to consider research/action methods which help to understand why farmers manage their activities the way they do, why they choose particular crops or livestock activities, and what changes might improve their lives.

Step 1: Understand the broader environment

Go to a village and learn as much as you can about it:

- What resources does it have (land, labour, climate, water, processing equipment, road access, buses, public services such as health, extension, police, school, etc.)?
- How are these resources used (in agriculture, livestock, forestry, mining, etc.)?
- What are the dynamics of the village? (Are people coming into the area, or are more people leaving?) If people are leaving, why? What crops are growing in importance?
- What are the causes of any conflicts? What are the scarce resources?

Step 2: Understand the types of farmers in that area

Talk to different farmers to find out about:

The resources they can mobilise

1. Land
   - The sizes of holdings and whether there is more land available for those who need it.
   - The quality of the land (its fertility, access to water, steep slopes, scattered holdings, conflicts, etc.)

- Utilisation of the land (which crops or livestock activities are important? Is land rented out to others? Is some land left fallow or resting?)
2. Family / Labour
   - How many people in the family contribute labour and depend on that land for food and money? Do they all work every day?
   - Do some have specialist knowledge or skills (carpenter, nurse, teacher, etc.)?
   - Do they sometimes work for others?
   - Do they sometimes depend on external labour?
3. Capital
   - What are the assets available (productive equipment, transport, housing, etc.)?
   - What is the farm's capacity to mobilise cash?
4. Social
   - Systems of social support – can they depend on help from others in the village?
   - Are they members of associations or cooperatives?
   - Can they access more land if they want it?

The objectives they aim at

1. Risk adversity, gender equity, the future lives of their young people.
   - What are the farmers' objectives? Individually? As a family?
   - Are the farmers able to formulate these objectives, and to create strategies to attain them?
   - Are these objectives compatible with the resources available to them currently?
2. Agriculture as a choice?
   - Do the farmers see themselves as still being farmers in ten years' time? Do they see their children as farmers in the future?
   - If not, where do they see themselves after ten years? Where do they see their children in the future?

Operations / activities

1. Production systems used
   - What are the cropping/livestock systems?
   - What are the objectives of these systems?
   - What are the practices / tools / inputs / labour / etc. used in these systems?

- What is the timing of each operation in these production systems?
2. Marketing systems used
   a) What marketing system is used for each crop?
3. Are there other ways in which they can market crops?
4. How much food does the family take from the farm to sustain its needs?

Calculations of profit from each crop

1. Calculate:
   - the gross margin from each crop (the amount by which the income from the crop exceeds the expenditure incurred in growing it).
   - the gross margin for each crop as a % of the income.
   - the total gross margin from all the crop and livestock activities.
   - the gross margin per unit of land (total gross margin for one hectare used to grow each crop).
   - the gross margin for each crop per unit of family labour (gross margin / amount of family labour – expressed in person-days – used to generate that profit).
2. Discuss the results with individual farmers or groups of farmers:
   - Compare the outcomes for each crop.
   - Compare the outcomes for different crops or combinations of crops.
   - Discuss the options most suited to farmers with limited access to land, limited access to labour and limited access to cash / working capital.
   - What is the sustainability of these systems? Do they produce enough to satisfy the farm and family needs; and to allow its replication over the years? What would happen if some parameters change (costs, outputs, selling prices, family needs, etc.)?
   - What alternatives could farmers consider to achieve realistic improved outcomes?

The farmers can then draw their own conclusions (if possible comparing their own calculations with those of other farmers, to discover which systems perform better and find out how the other farmers handle the system in terms of practices, inputs used, etc.).

Step 3: Segmentation and benchmarking

This will shed new light on the different types of farmers in that area, and how they inter-relate with each other and with the environment. It will also show the different scales of farms in the area, and the criteria that distinguish the different farmers and farms.

So each farmer can think about whether the conclusions apply to them, or only to some other farmers who have different circumstances.

The farmers can also get a feel of the possibilities for innovation and improvement - the capacity of the farming system to support changes that could lead to better livelihood (wealth, comfort, investment capacity, satisfaction of members, perspectives for the children to pursue the same activity, etc.).

Step 4: Training / Transition towards "management advisory services" / farmers' autonomy in self-assessing and deciding about the options they want to use

The key is to bring together research and action. Farmers certainly need to adapt to changing circumstances, but to drive such change, the innovations must fit with their own motivations, capabilities and understandings of how they can proceed.

So anyone who works with small farmers must find ways to motivate them, to make them feel more comfortable about their environment, less fearful of risks. To let them choose what and how and when, but give them all support when they need it to succeed (essentially understanding and conducive environment). Once they are motivated, the real challenge is to supply, in a coordinated way, all the "public goods" needed (at the same time) for them to succeed (advisory services, inputs, infrastructures, services, finance, markets, information, stable environment, etc.) To give them tools so that they can:

1. self-assess,
2. identify and assess solutions that are realistic and feasible for them,
3. work out a plan for each month
4. monitor their own performance.

INDEX

Accreditation schemes (for agricultural products), 136, 137, 146-51, 158-160, 223
Acidity (of soils), 12, 14
Adaptation, to climate change, 22-4, 25. *See also* climate change.
African Artemisia Ltd, 185, 188
AGRA (Alliance for a Green Revolution in Africa), 213, 256
Agribusiness, 118-9, 120, 169-170
Agricultural credit, 106, 109, 164-168, 172, 173-4, 255, 256
  for cotton, 177-8, 179-180
  for tobacco, 178-9, 179-180
Agricultural machinery, *see* Machinery, agricultural
Agricultural Research, 71-92, 199-201, 204-5, 256-7
  agricultural economics, 82-3, 101-5
  agronomic trials, 76-7
  control of pests and diseases, 16-19, 81-2
  interests of international companies, 83-4
  plant breeding, 10, 77-79
  animal breeding, 10, 79-80
  seed multiplication, 79-81
Agricultural extension, 85, 100-1, 111-2, 172, 192-202, 204-5, 237, 252, 257-258

Agricultural policies, 245-259
Alternative fuels, for local use, 253
Animal diseases, 16-9, 78, 119, 215-7, 218-9, 256
Antibiotic use and resistance, 118, 119, 218-9
Arable land, 107, 121, 130
Artemisinin, *Artemisia annua*, 168, 185-188
Auctions, 138, 141, 142, 170
Bacteria, 11, 12, 17, 81, 215, 218
Bacterial contamination, 146
Bananas, 17, 37-38, 79, 81, 85, 96, 218
Beans, 37, 38, 85 *See also* Legumes
Bibby, Andrew, 152-7
Big Results Now, 125-126
Biodiversity, 56, 80, 148, 217-218, 222
Biological control, 18, 217, 256
Boreholes, 43, 52, 58-9
Boserup, Ester, 45-6
BRC certification, *see* Accreditation schemes
Breeding of crops and animals, *see* Agricultural Research
Brooke Bond, 118
Bryceson, Debbie Fahy, 100-101, 108, 113
Byerlee, Derek, 126
Cafedirect, 158-160
Carbon 10-11, 23

Carbon dioxide, 10-11, 19, 52

Cashew nuts, 29, 39, 140, 144, 145, 156, 174, 203, 215

Cassava, 10, 17, 18, 22, 30, 37,74 (footnote), 79, 81, 85, 86-88, 140, 204, 222, 254, 259

Certification of agricultural products, see Accreditation schemes

Chambers, Robert, 85, 199-200

Climate change (Global warming), 14-15, 19-21, 22-24, 44-46, 106, 147, 220-1, 246

Coffee, 40, 60, 154, 173

Coffee Berry Disease, 82, 215

Coffee processing, 145

Commonwealth Development Corporation (CDC), 118

Conservation agriculture, 33-35, 41, 49, 116, 130, 132, 223, 224, 248, 251

Continuous (permanent) cultivation, 16, 30, 45, 97, 124

Contract farming, 4, 118-9, 165, 168-173, 176, 177-178, 258
cotton, 177-178
tobacco, 178-179
dairy, 181-184
Artemisia, 185-188
advantages and disadvantages, 171-173

Co-operatives, 99, 142-3, 151, 152-157, 171, 175, 177, 178-9, 255

Co-operative Act (Tanzania), 142

Co-operative farms, 128-133

Co-operative Reform and Modernisation Programme, 153-157

Co-operative Unions, 142-144, 177-179, 181-4

Corruption, 143, 156, 176

Cotton, 40, 82, 84, 103-4, 142, 143, 177-9, 217, 221

Coulson, Andrew, ix, 81, 84, 90, 113, 190, 229

Credit, see Agricultural credit

Crop diversification, 33, 97, 101, 131

Crop finance, 166-167

Crop prices, 2, 38, 40, 72, 87, 99, 106, 144, 146-8, 150, 173-5, 176, 177-9, 182-3, 203, 223-4, 225-6

Crop processing, 39, 85, 86-9, 116, 120, 121, 131, 136, 139-140, 143, 144-145, 203, 258

Crop storage, 85, 139, 140, 151, 253. See also Warehouse Receipt Schemes

Dairy farming, 41, 181-4

Dakawa Rice Farm, 67-68

Dams, 23, 53, 54, 58, 61, 66

Deininger, Klaus, 126, 134

Dimoto, Romanus, 241-243

Disputes (over land and water), 43, 68, 123, 251-2

DNA, 219-220

Drainage, 56-57, 66, 121

Drip irrigation, 60-61

Dwarf varieties, 80, 91, 203

Economies of scale, 115, 120-1, 128-133, 248, 258

Egziabher, Tewolde, 72

Ellman, Antony, ix, 128, 158, 185

Erosion control, 15-16, 20, 32, 61, 82, 97, 99, 107, 130-132, 224-6, 228, 241-2, 246

Exotic breeds, 42

Extension Officers, see Agricultural extension

Factory farming, 118-119, 121

Fair trade certification, see Accreditation schemes

Farmer groups, 125, 141-142, 166, 172, 176, 197, 255

Farming Systems Studies, 82-83, 103-105, 111-112, 234

Farm management studies, see Farming Systems Studies

Fast food revolution, 169

Female headed households, 235

Female extension workers, 252

Fixed assets, 167

Forage crops, 24, 40, 250

Fox, Bruce, 56, 69

Freire, Paulo, 198-199

Fruits, 39, 140-141.
  See also Horticulture
Fungi, 17, 19, 39, 40, 78, 81, 82, 215
Fungicides, 76, 204-5, 211, 215
Gatsby Trust. See Tanzania
  Gatsby Trust.
Gender Myths and Half Truths,
  230-240
Genes, genome, 10, 17, 77-80, 83,
  219-220
Genetic engineering, 220
Genetic Modification, see GM
Global warming, see Climate Change
GlobalGAP certification, 147
Glyphosate, see Roundup
GM (Genetic Modification), 79, 148,
  211-224, 245, 249
GM maize, 35-36, 222
Gravity irrigation,
  see Irrigation, gravity
Green Revolutions, 35, 74 (footnote),
  80, 205, 211-213, 245, 257
Gross margin analysis, 82, 102-105,
  111-112
Groundnut Scheme, 118
Hand cultivation, 31-2, 34, 39, 40,
  216, 217
Herbicide, see Weedkiller
Hoe cultivation, see Hand cultivation
Horticulture, 2, 29, 38-39, 41, 101, 131,
  122, 145, 149, 169, 173, 222, 232, 259
Humus, 13-14, 28
Hybrid seeds, 29, 35, 36, 80-81, 83, 193,
  194, 213, 223-4, 249
IITA (International Institute for
  Tropical Agriculture), 18, 87-88
Indigenous Technical Knowledge, 22-3,
  72, 98, 101, 109, 198-202, 236,
  239, 246, 267-8
Industrial crop processing,
  see Crop processing
Innovation in agriculture, 45-46, 72-4,
  78, 83-84, 98, 105, 108, 150, 154,
  192-5,199-200, 201-202, 246, 247,
  256, 271
Intensification of production, 29, 119,
  129, 132-133, 203, 249

Intensive livestock production, 117-9,
  218, 250
Intercropping, see Multiple cropping
Investments outside agriculture, 252
Iraqw tribe, 128
Irrigation, 3-4, 20, 22-4, 29, 36, 39,
  51-64, 82, 106, 108, 121, 193, 212,
  235, 239, 251-2
  bottle, 61
  gravity, 54-8, 62
  small scale, social organisation
    of, 57-8
  drip, 55, 62-3, 66
  flood, 52-53
  large scale,
  myths about, 3,
  sprinkler, 62, 66
  economics of, 62-65
Irrigation, arrangements for
  distribution of water, 57-58
Ismani area, 246
Jayne, Thomas, 6
Kenya, 2, 29, 35-6, 38, 42, 54, 117-8,
  147, 185-7, 204-5
Kibbutz, 129
Kilcher, Frédéric, 4, 111-112, 260-271
Kikula, Idris, 45, 48
Knowledge, of farmers. See Indigenous
  Technical Knowledge
Kyela rice, 36
Land Acts (Tanzania), 124
Land grabbing, 125-126, 240
Land tenure, 123-125, 248-249
Land titles, 166, 248-249
Land types, 29-30
Land use, 28, 31-35, 41
Large and small farms, comparison,
  119-123, 126-127
Large farm categories, 116-119
Large farms, 116-127, 166ff., 215, 255
Larger grain borer, Prostephanus
  truncates, 81
Lema, Ninatubu, 90
Livestock, 41-44, 77-78, 93-96, 102-103,
  231-232, 236, 247, 250. See also
  Intensive livestock production

M-Pesa, 150, 166 (footnote), 236
Maasai tribe, 24, 128-129
Machinery, agricultural, 31-32, 130, 131, 132, 215, 251
Maize, 35-6, 78, 102, 175, 212-3
Manure, 11, 16, 29, 214
Market information, 149-151
Market regulation and inspection, 145-146, 254-255
Marketing Boards, 143-144
Marketing concept, 137-138, 254
Marketing functions, 139
Marketing institutions, 141-144
Marketing systems, 136-151
Matchmaker Associates, 173, 181
Mbiha, Emmanuel, x
Mechanisation, see machinery
Mdee, Anna, 67-8
Mediplant, Switzerland, 185
Minimum tillage, No-till, 33, 132, 223
Mixed farming, 41, 49, 250
Miyashita, Chie, 8, 223, 225
Mobile phones, 136, 149-150, 166 (footnote), 194
"Modern" agriculture, 108-109
Monocropping, 148, 251
Monsanto company, 73, 83, 212, 220 (footnote)
Moshav shitufi, 129
Movable assets, 167
M-Pesa, 150, 166 (footnote), 190, 236
Mrema, Ezra, 215, 228
Multiple cropping, 97-98, 119, 251
MVIWATA, 150
Myths about agriculture in Africa, 3, 106-7, 228. See also Gender Myths
Nerica (New Rice for Africa) rice, 36
Nitrogen, nitrogen fixation, 11, 17, 33, 37, 214, 251
Njuguna, Catherine, 86-87
Nucleus estates, 116
Nutrition (of people), 41, 95, 130, 248, 259
Nuts, 39
Opportunity costs, 102

Organic certification, see Accreditation schemes
Organic farming, 147, 148-9, 173, 214, 216-7, 222-224, 225-227, 249
Organic matter, 10-15, 33-34, 100
Outgrowers, 40, 116, 121, 169, 171, 196, 197
Overdraft facility, 167
Ox cultivation, 31, 49, 131, 240
Pan-territorial prices, 144
Pastoralism, 24, 42, 106-7, 250
Peasant resistance, 100-1
Peasants, 107-108
Pest and disease control, 16-19, 75, 77, 81-82, 88, 215-217, 256
Pests and diseases, 9, 16-19, 40, 47, 63, 77, 78, 81-2, 140, 215-7, 221, 256
Plant breeding, see Agricultural Research
Plantations or estates, 116-117
Plant growth, 10-21
Population growth, 44-46, 132-3, 236
Post harvest losses, 140
Potatoes (round, Irish), 38, 204-5
Potts, David, 190
Prices, see Crop prices
Primary Co-operative Societies, 142-143, 152-154, 168, 175, 177-180, 181-184,
Primary crop processing, 144-145
Private sector traders, 141
Processing, see Crop processing
Producer groups, 99, 141, 151, 166, 258
Protein, 10, 11, 27, 41, 78, 88-9, 217, 222, 250, 259
Public health, see Nutrition
Rainwater harvesting, 61
Ranches, 42, 117, 250
Rice (paddy), 36, 67-68, 79, 104, 122
Risk, risk analysis, 167, 251
Roundup™ (or Glyphosate), 212, 220-1
Rural infrastructure, 99
Rwanda, 213, 251
SACCOs, 152, 157
Salinity, 56, 63, 66

Scott, James, 100

Seed laws, 80-81

Seed multiplication, 79-80

Settlers, 117

Sewando, Ponsian, 89, 91

Shifting cultivation, 16, 26, 30, 31, 45

Side selling, 165, 172-173, 179-180

Sisal, 39-40, 116-117, 145, 171

Small farmer motives, 97-98, 109-110

Small farmer responses, 101-103, 106

Small farms, 1,-3, 4, 5, 6, 23, 31-2, 34, 47, 72, 84-5, 95-112, 193

Small-scale mining, 100-101

Small-scale irrigation, 24, 53-4, 57, 61, 66

Soil erosion, 15-16, 20, 32-33, 61, 82, 97, 99, 132, 245-246

Soil science, 82

Soils, 10-14, 20, 28-30, 36, 77, 82, 224, 246

Sorghum, 36

Spices, 39

Sprinkler irrigation, 59-60

State Farms, 118

Status Report on Poverty in Sub-Saharan Africa,231, 232, 233, 235

Subsidies, 112,113, 122-123, 255

Sugar, sugarcane, 62, 116, 121-122, 171, 235 (footnote), 259

Sulle, Emmanuel, 125, 134

Sunflower, 38, 145, 194

Supermarket chains, 169-170

Sweet potatoes, 38, 47

Tanga Dairy Co-operative Union, 181-4

Tanga Fresh Ltd, 181-184

Tanzania Gatsby Trust, 18, 168, 178

Tea, 40, 79, 118, 158-160

Teadirect, 158-160

Technology, 72-74, 251-252. See also Innovation.

Tobacco Industries Act, 142, 168

Tobacco, 40, 171, 178-180

Tractor cultivation, 32, 49, 129-131. See also Machinery, agricultural

Traidcraft plc and Traidcraft Exchange, 148

Training and Visit, 196-197, 199

Transformation strategy, 109, 128-129

Tree planting, 15-16, 20, 29, 30, 33, 121, 193, 197, 253. See also: Forests

Trickle irrigation, See irrigation, drip

Ujamaa Villages, 113, 128 (footnote)

Upper Kitete farm, 128-133

Urban agriculture, 61-62

UWAWAKUDA (Society of Small Farmers in Dakawa), 67

Uyole Agricultural Research Station, 193, 204

Value chains (value chain analysis), 86-7, 136, 137-140, 151

Vegetables, see Horticulture

Vegetative propagation, 10, 78-9, 81

Village Development Committees (Village Associations), 124, 130-132

Village Settlements, 128

Villagisation, 124, 131-133

Viruses, 17, 43, 81, 88, 215

Warehouse Receipt Schemes, 156, 164, 165, 173-175, 176

Water lifting devices, 53-54,

Weedkillers, 18-19, 34, 212, 220-1

Wegerif, Mark, 181

Wheat, 29, 36-37, 118, 120, 129-130, 212

Wilts, 17, 37-8, 85

Women and development, 230-240

Women farmers, 99, 252

Women, discrimination against, 235-236, 238-240, 245

Women's education, 236

Words, misuse of, 107-9

Working capital, 167

Wuyts, Marc, 144

Printed in the United States
By Bookmasters